张 亮 编著

Swift
从入门到精通

清华大学出版社
北京

内 容 简 介

本书基于最新的苹果官方 Swift 4 语法规范,所有的示例程序均在 Xcode 10 中开发完成。本书共分四篇,分别为基础篇、高级篇、深入篇和应用篇。其中,基础篇介绍最基本的语法概念,包括数据类型、控制流和函数等;高级篇讲解面向对象相关的知识,包括结构体、类和协议等;深入篇讨论相对复杂的语法特色,包括可选链、扩展和内存管理等;应用篇侧重分析如何运用前面学习的 Swift 知识解决实际问题,包括由易到难的两个应用实例。

本书不仅可作为信息类相关专业本科生的移动开发课程配套教材,也可作为 iOS 开发工程师和 iOS 开发兴趣爱好者的 Swift 语法参考书籍。

本书封面贴有清华大学出版社防伪标签,无标签者不得销售。
版权所有,侵权必究。举报: 010-62782989,beiqinquan@tup.tsinghua.edu.cn。

图书在版编目(CIP)数据

Swift 从入门到精通/张亮编著. —北京:清华大学出版社,2019.12(2024.7重印)
ISBN 978-7-302-54331-2

Ⅰ. ①S… Ⅱ. ①张… Ⅲ. ①程序语言-程序设计 Ⅳ. ①TP312

中国版本图书馆 CIP 数据核字(2019)第 259436 号

责任编辑:张瑞庆　常建丽
封面设计:常雪影
责任校对:李建庄
责任印制:曹婉颖

出版发行:清华大学出版社
网　　址:https://www.tup.com.cn,https://www.wqxuetang.com
地　　址:北京清华大学学研大厦 A 座　　邮　编:100084
社 总 机:010-83470000　　邮　购:010-62786544
投稿与读者服务:010-62776969,c-service@tup.tsinghua.edu.cn
质量反馈:010-62772015,zhiliang@tup.tsinghua.edu.cn
课件下载:https://www.tup.com.cn,010-83470236
印 装 者:三河市铭诚印务有限公司
经　　销:全国新华书店
开　　本:185mm×230mm　　印　张:22.5　　字　数:385 千字
版　　次:2019 年 12 月第 1 版　　印　次:2024 年 7 月第 5 次印刷
定　　价:69.00 元

产品编号:084429-01

前言

本书全面介绍了 Swift 4 中的所有重要特色,但又摒弃了部分不适合初学者的冷僻知识点,使第一次接触 Swift 的读者能够抓住重点,掌握精髓。除了在 Swift 语言方面尽最大可能做到面面俱到,本书通过预备知识部分,帮助读者迅速掌握 Xcode 的基本功能,熟练运用 playground 编写代码,为配合后面章节中编写小实例打下必要的基础。介绍完 Swift 语言,紧随其后的是由浅入深、逐步分解问题的一个应用实例,帮助读者运用已学的 Swift 语言解决实际问题,使读者体会到 Swift 语言在实际 iOS App 开发中扮演的角色。最后部分通过 30 个比较经典的编程练习帮助读者巩固已学的 Swift 语言知识。

本书共分为 4 篇。

第 1 篇为基础篇,含 8 章,介绍了预备知识和 Swift 语法中的基础内容。其中,第 1 章介绍读者在阅读本书 Swift 语言知识前必须掌握的基础内容,内容主要包括 Xcode 特性、Xcode 的使用以及 playground 介绍。playground 将是 Swift 语言实践的主要工具。第 2 章介绍基本的数据类型,包括常量和变量、整型和浮点型、布尔型、元组型、可选型等内容。第 3 章介绍各种常用的运算符,包括赋值运算符、算术运算符、关系运算符、逻辑运算符、三元运算符、区间运算符。第 4 章介绍字符串的相关知识,包括字符串的定义、字符串的操作等。第 5 章介绍几种集合类型的定义和操作,包括数组、集合和字典。第 6 章介绍几种控制流的语法和使用方法,包括 for 循环、while 循环、if 条件语句、switch 条件语句及控制转移语句。第 7 章主要介绍函数的相关知识,包括函数的定义和调用方法、函数形参、函数类型、嵌套函数。第 8 章主要介绍闭

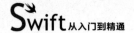

包的相关知识，包括闭包表达式和闭包的应用。

第 2 篇为高级篇，含 7 章，介绍了 Swift 语法中的面向对象的基础知识。其中，第 9 章介绍枚举型，包括枚举型的定义、枚举型的关联值及原始值的定义和用法。第 10 章介绍结构体与类的定义，比较了两者的共同点和不同点，以及各自的应用场景。第 11 章介绍属性的相关知识，包括存储属性、计算属性、属性观察器以及类型属性。第 12 章介绍方法的相关知识，重点介绍实例方法和类型方法，同时也介绍了比较特殊的下标方法。第 13 章介绍类，先介绍了基类和子类的概念，然后介绍了如何进行重载。第 14 章介绍了协议，包括协议的使用、协议的继承性。第 15 章介绍了泛型的相关知识，包括泛型函数、泛型类。

第 3 篇为深入篇，含 7 章，介绍 Swift 语法中的面向对象的高级知识。其中，第 16 章介绍异常处理的相关知识，包括异常的定义及如何进行异常处理。第 17 章介绍了可选链，包括强制拆包的概念，以及可选链的应用场景。第 18 章介绍访问控制的相关定义和应用场景。第 19 章介绍类型操作符的相关知识，包括类型检查和类型转换。第 20 章介绍扩展的概念，包括扩展计算型属性、扩展构造器、扩展方法及扩展下标。第 21 章介绍内存管理的相关知识，包括内存管理的工作原理、强引用循环的概念及消除的方法、闭包中的强引用循环等。第 22 章介绍高级运算符，包括位运算符、溢出运算符及运算符函数。

第 4 篇为应用篇，含 4 章，通过实现两个完整的苹果应用，在实际工程项目中综合运用前面学习的 Swift 语言知识。第 23 章以计算器应用为例简要介绍了苹果应用的相关知识，包括如何建立一个简单的应用、iOS App 的 MVC 架构以及应用的运行状态的切换。第 24 章介绍了如何编程实现计算器应用，包括界面设计、动作处理和运算逻辑等部分。第 25 章介绍如何基于 MVC 架构实现手游 2048 的过程，分别从模型层、视图层和控制层分析和实现。第 26 章为 Swift 语言的编程练习及参考答案，通过 30 个比较经典的编程练习题，讲解每个题的编程思路，并给出代码、相关代码说明及系统运行结果。

由于编者水平有限，有不少考虑不周的地方和不足之处，敬请使用本书的老师、同学及广大读者批评指正。

编　者

2019 年 7 月

第1篇 基 础 篇

第1章 预备知识 / 3

1.1 Xcode 特性 …………………………………………… 3
1.2 Xcode 的使用 …………………………………………… 4
1.3 playground 介绍 ………………………………………… 7

第2章 数据类型 / 10

2.1 常量和变量 …………………………………………… 10
2.2 整型和浮点型 ………………………………………… 16
2.3 布尔型 ………………………………………………… 19

2.4 元组型 …………………………………………………………… 20
2.5 可选型 …………………………………………………………… 22
本章知识点 ………………………………………………………… 29
参考代码 …………………………………………………………… 30

第 3 章　运算符 / 35

3.1 赋值运算符 ……………………………………………………… 35
3.2 算术运算符 ……………………………………………………… 36
3.3 关系运算符 ……………………………………………………… 37
3.4 逻辑运算符 ……………………………………………………… 38
3.5 三元运算符 ……………………………………………………… 40
3.6 区间运算符 ……………………………………………………… 41
本章知识点 ………………………………………………………… 41

第 4 章　字符串 / 42

4.1 字符串的定义 …………………………………………………… 42
4.2 字符串的操作 …………………………………………………… 44
本章知识点 ………………………………………………………… 50
参考代码 …………………………………………………………… 50

第 5 章　集合 / 54

5.1 数组 ……………………………………………………………… 54

5.2 集合 …………………………………………………………… 60
5.3 字典 …………………………………………………………… 63
本章知识点 ………………………………………………………… 66
参考代码 …………………………………………………………… 67

第 6 章　控制流 / 72

6.1 for 循环 ………………………………………………………… 72
6.2 while 循环 ……………………………………………………… 73
6.3 if 条件语句 ……………………………………………………… 74
6.4 switch 条件语句 ………………………………………………… 75
6.5 控制转移语句 …………………………………………………… 78
本章知识点 ………………………………………………………… 80
参考代码 …………………………………………………………… 80

第 7 章　函数 / 84

7.1 函数的定义和调用方法 ………………………………………… 84
7.2 函数形参 ………………………………………………………… 86
7.3 函数类型 ………………………………………………………… 89
7.4 嵌套函数 ………………………………………………………… 92
本章知识点 ………………………………………………………… 93
参考代码 …………………………………………………………… 93

第 8 章 闭包 / 98

8.1 闭包表达式 …………………………………………………… 98
8.2 闭包的应用 …………………………………………………… 103
本章知识点 ………………………………………………………… 106
参考代码 …………………………………………………………… 107

第 2 篇 高 级 篇

第 9 章 枚举型 / 111

9.1 枚举型的定义 ………………………………………………… 111
9.2 关联值 ………………………………………………………… 114
本章知识点 ………………………………………………………… 117
参考代码 …………………………………………………………… 118

第 10 章 结构体与类 / 120

10.1 基本概念 ……………………………………………………… 120
10.2 值类型与引用类型 …………………………………………… 123
本章知识点 ………………………………………………………… 126
参考代码 …………………………………………………………… 126

第 11 章　属性 / 128

11.1　存储属性 ·· 128
11.2　计算属性 ·· 130
11.3　属性观察器 ······································· 133
11.4　类型属性 ·· 135
本章知识点 ·· 136
参考代码 ··· 136

第 12 章　方法 / 139

12.1　实例方法 ·· 139
12.2　类型方法 ·· 141
12.3　可变方法 ·· 142
12.4　下标方法 ·· 142
本章知识点 ·· 145
参考代码 ··· 145

第 13 章　类 / 150

13.1　继承性 ··· 150
13.2　重载 ·· 155
13.3　类的构造 ·· 158
13.4　类的析构 ·· 166

本章知识点 ·· 168

参考代码 ·· 168

第 14 章　协议 / 175

14.1　协议的使用 ··· 175

14.2　协议的继承性 ·· 181

本章知识点 ·· 183

参考代码 ·· 183

第 15 章　泛型 / 189

15.1　泛型函数 ·· 189

15.2　泛型类 ··· 191

本章知识点 ·· 192

参考代码 ·· 192

第 3 篇　深　入　篇

第 16 章　异常 / 197

16.1　定义异常 ·· 197

16.2　抛出异常 ·· 197

16.3　处理异常 ·· 199

本章知识点 …………………………………………………………… 200

参考代码 ……………………………………………………………… 201

第 17 章　可选链 / 202

17.1　可选链的定义 ……………………………………………………… 202

17.2　可选链的使用 ……………………………………………………… 203

本章知识点 …………………………………………………………… 207

参考代码 ……………………………………………………………… 207

第 18 章　访问控制 / 210

18.1　访问级别 …………………………………………………………… 210

18.2　实例 ………………………………………………………………… 211

本章知识点 …………………………………………………………… 214

参考代码 ……………………………………………………………… 214

第 19 章　类型操作符 / 216

19.1　类型检查 …………………………………………………………… 217

19.2　类型转换 …………………………………………………………… 218

本章知识点 …………………………………………………………… 219

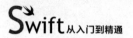

第 20 章　扩展 / 220

20.1　扩展计算型属性 ·········· 220
20.2　扩展构造器 ·········· 221
20.3　扩展方法 ·········· 222
20.4　扩展下标 ·········· 222
本章知识点 ·········· 223

第 21 章　内存管理 / 224

21.1　工作原理 ·········· 224
21.2　强引用循环 ·········· 225
21.3　闭包中的强引用循环 ·········· 230
本章知识点 ·········· 232

第 22 章　高级运算符 / 233

22.1　位运算符 ·········· 233
22.2　溢出运算符 ·········· 235
22.3　运算符函数 ·········· 235
本章知识点 ·········· 238

第 4 篇 应 用 篇

第 23 章 苹果应用 / 241

23.1 一个简单的应用 …………………………………………………… 241
23.2 MVC 架构 …………………………………………………………… 250
23.3 应用运行状态 ………………………………………………………… 252

第 24 章 计算器 / 254

24.1 界面设计 ……………………………………………………………… 254
24.2 动作处理 ……………………………………………………………… 259
24.3 运算逻辑 ……………………………………………………………… 266
24.4 小结 …………………………………………………………………… 276

第 25 章 手游 2048 / 277

25.1 模型层 ………………………………………………………………… 278
25.2 视图层 ………………………………………………………………… 288
25.3 控制层 ………………………………………………………………… 296
25.4 小结 …………………………………………………………………… 305

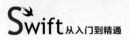

第26章　编程练习及参考答案 / 306

【练习1】　Fibonacci 数列 …………………………………………………… 306
【练习2】　求质数 ………………………………………………………… 307
【练习3】　求水仙花数 …………………………………………………… 309
【练习4】　统计字符串中的各类字符个数 ……………………………… 310
【练习5】　给定项数和数的和 …………………………………………… 311
【练习6】　自由落体反弹问题 …………………………………………… 312
【练习7】　求无重复的三位数 …………………………………………… 312
【练习8】　阶梯奖金计算 ………………………………………………… 314
【练习9】　求完全平方数 ………………………………………………… 315
【练习10】　求一年中的第几天 …………………………………………… 316
【练习11】　3 个数比大小 ………………………………………………… 317
【练习12】　打印九九乘法表 ……………………………………………… 319
【练习13】　猴子吃桃问题 ………………………………………………… 320
【练习14】　求分数数列的和 ……………………………………………… 321
【练习15】　求 n 的阶乘的和 ……………………………………………… 322
【练习16】　用递归法求阶乘 ……………………………………………… 322
【练习17】　倒推年龄 ……………………………………………………… 323
【练习18】　逆序打印一个整数 …………………………………………… 324
【练习19】　回文问题 ……………………………………………………… 325
【练习20】　整数排序 ……………………………………………………… 327
【练习21】　求 5×5 矩阵对角线之和 …………………………………… 328
【练习22】　折半查找 ……………………………………………………… 329
【练习23】　围圈报数 ……………………………………………………… 330
【练习24】　求分数的和 …………………………………………………… 332
【练习25】　字符串排序 …………………………………………………… 333

【练习 26】 猴子分桃问题……………………………………………… 334
【练习 27】 考试成绩统计………………………………………………… 336
【练习 28】 子串出现的次数……………………………………………… 338
【练习 29】 数字加密问题………………………………………………… 339
【练习 30】 被 9 整除问题………………………………………………… 340

第 1 篇

基 础 篇

基础篇由 8 个章节构成,分别为预备知识、数据类型、运算符、字符串、集合、控制流、函数以及闭包。这些是程序设计中的基础知识,其中大部分与其他编程语言类似,如整型和字符串、常量和变量以及控制流和函数。学习这部分内容时,读者要善于对比 Swift 语言和其他编程语言(如 C++ 和 Java),从相似中发现差别。另外,本篇还会介绍 Swift 语言中独有的内容,如可选型和闭包。对于这部分内容,读者除了要掌握其概念和正确地应用,还要深入思考向 Swift 语言中引入这些独特内容的原因。

第 1 章 预备知识

1.1 Xcode 特性

Xcode 是苹果公司发布的集成开发环境,可以在 Xcode 上开发 iPad、iPhone、Apple Watch 和 Mac 上的应用软件。Xcode 提供了一整套开发工具支持应用的全生命周期的开发,从应用创建到产品测试到系统优化,直到产品发布。

如图 1.1 所示,Xcode 集成了代码编辑器、用户界面设计、测试、调试等工具在一个用户界面中。用户可以根据自己的使用习惯和主要使用的功能自定义 Xcode 开发界面。

Xcode 提供了源代码辅助编辑功能,它不仅能对输入的源代码进行即时的语法和逻辑检查,提示错误,甚至提供修复错误的方案,而且还能根据少量的输入联想出可能的后续输入选项,帮助人们只输入很少的几个字母,就可以完成一段完整代码。

Xcode 的界面编辑器是一个可视化的编辑器,可视编辑器提供了一个组件库供直接使用。通过组合各种视窗和控件可以快速搭建基于 iOS、watch OS、OS X 的应用软件界面。特别值得一提的是,Xcode 中的 Storyboards 技术使开发者从繁杂的界面设计工作中解放出来,为开发者完成了大部分的界面设置和跳转工作,开发者可以更专注于业务流程的设计和代码的编写中。

自动布局是 Xcode 中的另一个重要特性。开发者可以定义界面元素的约束集,从而使开发出的界面可以适应各种屏幕尺寸和屏幕的不同方向显示。

Xcode 提供了强大的应用调试功能,不仅可以在模拟器上进行调试,也可以和硬件外设连接进行调试,对于开发人员来说,两者没有任何差别。

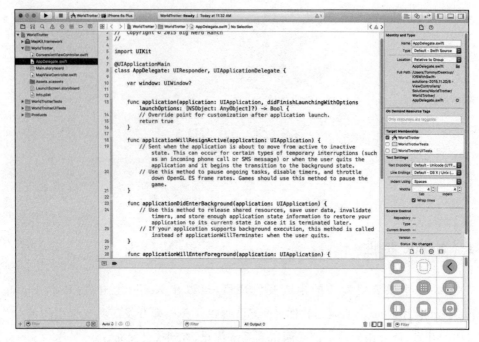

图 1.1 Xcode 开发界面

另外，Xcode 还提供了测试框架，帮助开发者快速建立功能测试、性能测试以及用户界面测试模块，使测试过程非常方便。

编写代码的时候，我们常常会查阅官方文档。Xcode 中集成了各种详细的官方技术文档，在 Xcode 开发环境中可以通过快速帮助很方便地找到需要了解的技术内容。

1.2 Xcode 的使用

Xcode 是免费软件，它运行在 Mac OS X 操作系统上。所以，要下载和运行 Xcode，首先需要有一台安装了 Mac OS X 的计算机，然后可以通过 App Store 下载。如图 1.2 所示，输入关键字 Xcode，单击"搜索"即可。

如图 1.3 所示，在搜索结果中找到 Xcode，单击 GET，然后单击 Install App 即可，本书安装的是 Xcode 10。

如图 1.4 所示，如果想卸载 Xcode，到 Launchpad 中找到 Xcode，直接拖入垃圾桶即可。

第 1 章 预备知识 Chapter 1

图 1.2 搜索 Xcode

图 1.3 下载 Xcode

图 1.4　卸载 Xcode

安装好 Xcode 后，打开 Xcode，如图 1.5 所示，标示为 1 的区域为工具栏，提供了最主要的操作快捷键。标示为 2 的区域为导航区，用来选择工作空间中的内容显示。

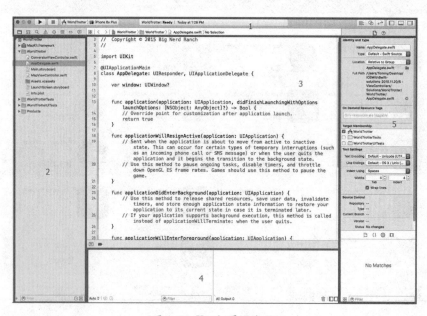

图 1.5　Xcode 界面分区

标示为 3 的区域为编辑区,如果在导航区选择的是一个源代码文件,则为代码编辑区;如果在导航区选择的是一个界面文件,则为界面编辑区。标示为 4 的区域为调试信息和系统输出信息显示窗口。标示为 5 的区域为组件区,在这个区域可以进行各种属性的设置工作,也可以选择图形元素并拖放到界面编辑区。

1.3　playground 介绍

在 Xcode 中编写代码有两种方式:一种是通过建立工程编写各种源文件,目的是发布一个应用;另一种是建立一个在 playground 中运行的文件,目的是为了学习语法和尝试代码的各种运行结果和效果。本书以讲解 Swift 语法为主,所以 playground 是最理想的编写代码和调试的工具,本书后面章节中的程序大部分都运行于 playground 环境。通过这种方式,可以让读者更加专注于学习 Swift 语法本身,而不用考虑创建工程相关的复杂过程。

下面来看怎么创建一个 playground 文件。首先,打开 Xcode,在弹出的启动页面中选择 Get started with a playground,如图 1.6 所示。

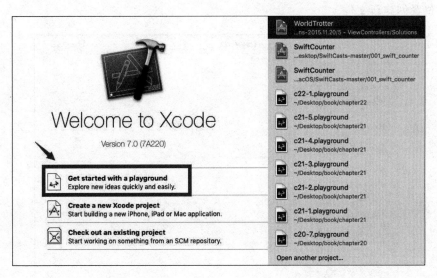

图 1.6　创建 playground 文件

如图 1.7 所示,在弹出的页面中输入文件的名字和运行的平台,这里默认输入 MyPlayground,平台默认选择 iOS,单击 Next 按钮。

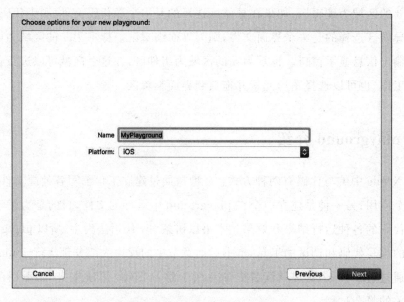

图 1.7　输入 playground 文件名

如图 1.8 所示，选择保存源代码的位置，这里选择默认位置，单击 Create 按钮，在默认位置创建新文件。

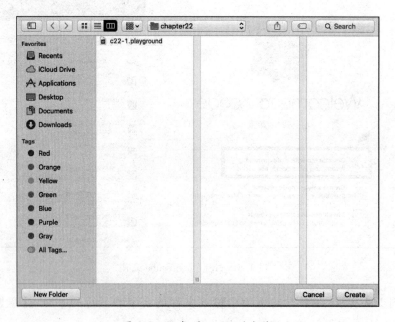

图 1.8　保存 playground 文件

图 1.9 为 playground 的编辑页面,默认会有一行代码,当编写代码时可以将其删除。在 playground 界面中,标示为 1 的区域为代码输入区,可以将要试验的代码写在这个区域。标示为 2 的区域为系统同步输出区,写完一行代码后,系统会自动运行,并将运行结果同步显示在对应的代码后面。如图 1.9 所示,给变量 str 赋值为一个字符串"Hello,playground",就会在这个区域的同一行显示 str 的值。标示为 3 的区域为调试信息区,这个区域是标准的系统输出区,如打印一个字符串,就会显示在这个区域。标示为 4 的区域为屏幕各种显示方式的快捷切换键。

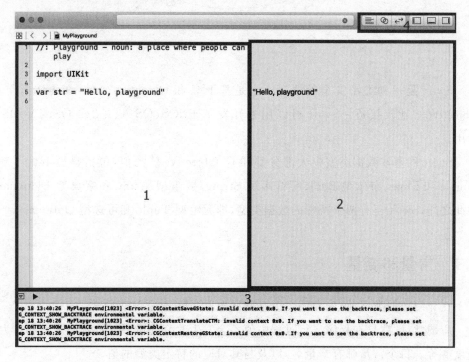

图 1.9 playground 界面分区

第 2 章 数据类型

Swift 是一种类型安全的语言。它是基于 C 和 Objective-C 的，在语法上与 C、Objective-C 比较接近。Swift 可以用来开发基于 iOS、OS X、watch OS、TV OS 的应用。

Swift 的基本数据类型中大部分都和 C、Objective-C 类似，包括整型 Int、浮点型 Double 和 Float、布尔型 Bool、字符串型 String、数组型 Array 和字典型 Dictionary。除此之外，Swift 还有两个特殊的数据类型，即元组型 Tuple 和可选型 Optional。

2.1 常量和变量

在介绍数据类型前，先介绍 3 个基础概念：常量、变量和表达式。

所谓常量，就是在程序运行过程中值不会发生改变的量，例如，圆周率 π 和重力加速度 g 等。每个常量都有常量名，以及与其对应的特定类型的值。

常量要用关键字 let 声明。常量的定义如图 2.1 所示。

```
3  let lengthOfTelephoneNumber = 8            8
4  let lengthOfIdentityCardNumber = 18        18
5  let lengthOfZipNumber = 6                  6
```

图 2.1　常量的定义

上例中定义了 3 个常量，分别为：固定电话号码的位数 lengthOfTelephoneNumber、身份证号码的长度 lengthOfIdentityCardNumber 以及邮政编码的位数 lengthOfZipNumber。

定义常量的格式为

let constantName=value

所谓变量,就是在程序运行过程中值可以发生改变的量,例如每天的温度和汽车行驶的速度。每个变量都有变量名,以及与其对应的特定类型的值。

变量要用关键字 var 声明。变量的定义如图 2.2 所示。

```
7  var totalVisits = 2000000
8  var numberOfMembers = 10000
9  var currentSpeed = 120
```

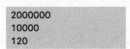

图 2.2　变量的定义

上例中,用变量分别定义了网站总访问量 totalVisits、会员数 numberOfMembers 以及汽车的当前速度 currentSpeed。

定义变量的格式为

var variableName=value

2.1.1　常量和变量的命名规则

命名常量和变量时,除了不能包含数学符号、箭头、保留的 Unicode 码位、连线和制表位,不能以数字开头外,没有任何其他限制。如图 2.3 所示,前 3 个语句的命名中分别使用了数学符号、连线和以数字开头,所以系统报错,后 3 个语句符合命名的规范,顺利通过编译。

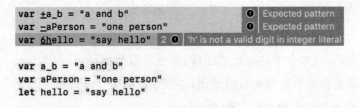

图 2.3　常量和变量的命名

2.1.2　有意义的名字

常量和变量的名字只要符合命名规则,就是合法的,就能通过编译,并被计算机正确理解。为了增加代码的可读性,还需要为常量和变量起一个易于理解的名字。

图 2.4 定义了两个变量和一个常量。通过它们的名字,很容易知道其代表的含义,因此本例中的命名是一个好的命名。

```
17  var numberOfStudents = 68                    68
18  var idCardNumber = "3206022016121120652"     "3206022016121120652"
19  let pi = 3.141516                            3.141516
```

图 2.4 有意义的名字

图 2.5 也定义了两个变量和一个常量,并且是符合命名规则的合法命名,但是可读性很差。即使写这段代码的程序员,用不了几天也会猜不出自己定义的变量和常量的含义。

```
21  var a = 68                                   68
22  var b = "3206022016121120652"                "3206022016121120652"
23  let c = 3.141516                             3.141516
```

图 2.5 可读性差的名字

2.1.3 驼峰拼写法

一般来说,有意义的命名往往是由多个单词组成的复合词。在不使用空格的情况下,有多种写法帮助程序员读出复合词中不同的单词。这里介绍一种驼峰拼写法(camel case),其也是苹果公司官方文档中使用的命名法,建议 Swift 程序员都采用和苹果公司官方文档中一致的驼峰拼写法。

苹果公司的驼峰拼写法要求遵守以下 3 条原则。

(1) 复合词的第一个单词的首字母小写,其他单词的首字母大写。

(2) 复合词中,每个单词除了首字母,其他字母都小写。

(3) 如果复合词中含有缩写的单词,则该单词的每个字母都大写。

图 2.4 给出的例子就采用了驼峰拼写法。

2.1.4 类型声明

定义常量或变量时,可以加上类型声明,告诉编译器该常量或变量中要存储的值的类型,同时还要进行初始化,即给该常量或变量赋初值。这一点是 Swift 4 新提出来的，以前的版本中并不要求在类型声明的同时进行初始化。

常量的类型声明格式为

let constantName : type=initialValue

变量的类型声明格式为

var variableName : type=initialValue

图 2.6 定义了一个 String 类型的常量 name 和一个 Int 类型的变量 age。

```
25  let name : String = ""
26  var age : Int = 0
```

""
0

图 2.6　类型声明

定义常量或变量时，可以不显式地声明其类型。当第一次为常量或变量赋值时，编译器通过推断得到其类型。

常量或者变量在声明时，如果声明了具体类型，就不能改变。常量与变量之间不能互相转换。也就是说，常量不能变为变量，变量也不能改为常量。常量的值一旦确定，就不能改变，而变量的值可以改变为同类型的其他值。如图 2.7 所示，当试图改变常量 hello 的值时，系统提示译成中文大意为"hello 为常量，不能对其赋值，通过将关键字 let 改为 var 修复这个问题"，而改变变量 a_b 的值是合法的。

```
13  var a_b = "a and b"
14  var aPerson = "one person"
15  let hello = "say hello"
16
17  hello = "say hi"
18  a_b = "c and d"
19
20
```

Cannot assign to value: 'hello' is a 'let' constant
Change 'let' to 'var' to make it mutable   Fix

图 2.7　类型声明错误

单击 Fix 按钮可以自动修复问题。如图 2.8 所示，系统不再提示错误。

```
13  var a_b = "a and b"
14  var aPerson = "one person"
15  var hello = "say hello"
16
17  hello = "say hi"
18  a_b = "c and d"
```

"a and b"
"one person"
"say hello"

"say hi"
"c and d"

图 2.8　修复类型声明错误

2.1.5 打印输出函数 print()

通过控制台输出程序执行过程中的运行结果，对于调试程序是非常有用的。Swift 4 提供了全局函数 print() 输出变量或常量的值到控制台 console 中显示。

调用格式为

print(varName)

也可以用常量名或变量名做占位符插入字符串中输出，输出结果将用当前常量或变量的值替换这些占位符。

调用格式为

print("string \(constName_or_varName)")

如图 2.9 所示，屏幕的底部区域为控制台输出区。

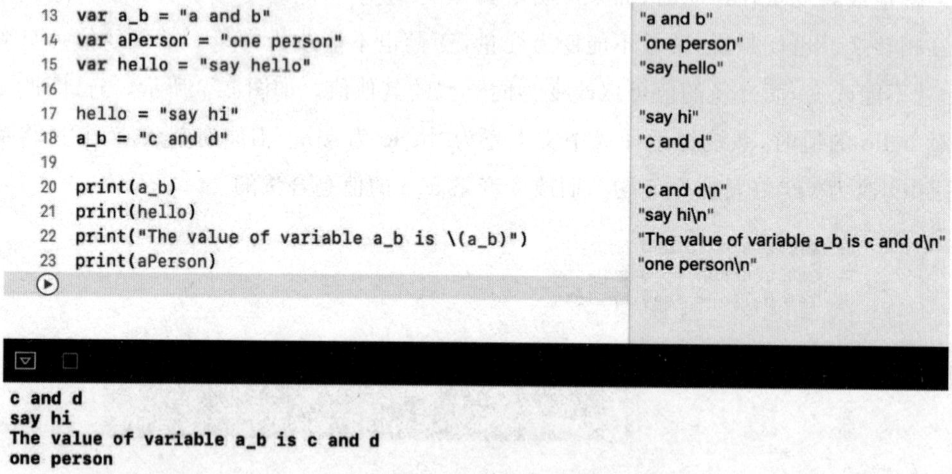

图 2.9　打印输出函数

2.1.6 向代码中添加注释

注释是代码的一个重要组成部分，它可以大幅提升代码的可读性，对于代码交接和维护来说至关重要。一般高质量的代码中大约有三分之一的篇幅为注释。

和 C 语言一样，Swift 也提供了两种形式的注释：一种是以双斜杠（"//"）为起始的单行注释；另一种是以斜杠加星号("/*")为起始、星号加斜杠("*/")为终止的多行注释。

如图 2.10 所示，向代码中添加注释。

```
1  import UIKit
2  /*
3   This file is designed for introducing how to use comments
4   Author: Liang Zhang
5   Organisation: BUAA
6   Created date: Jan 1st, 2019
7  */
8  
9  //definition of var
10 var a_b = "a and b"
11 var aPerson = "one person"
12 var hello = "say hello"
13 
14 //assignment of var
15 hello = "say hi"
16 a_b = "c and d"
17 
18 //output the value of var
19 print(a_b)
20 print(hello)
21 print("The value of variable a_b is \(a_b)")
22 print(aPerson)
```

图 2.10　添加注释

2.1.7　语句分隔符

编写 Swift 程序时，写完一行后，一般不需要加分号作为结束标识，可以直接按 Enter 键并开始写第二行语句，非常便捷。当要在一行中输入多条语句时，可以使用逗号作为一条语句的分隔符。如图 2.11 所示，通过逗号分隔 3 个变量赋值语句。每行一条常量赋值语句，语句间通过换行符分隔。

```
3  var a_b = "a and b", aPerson = "one person", hello = "say hello"
4  
5  let thisYear = "2019"                             "2019"
6  let myOrganisation = "BUAA"                       "BUAA"
7  let myhobby = "Reading"                           "Reading"
```

图 2.11　语句分隔符

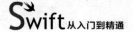

Swift是类型安全语言,因此编译器在编译代码的时候会进行类型检查,并将类型不匹配的语句用报错的方式提示程序员。声明常量和变量时,如果没有显式指定数据类型,Swift就会根据对其的赋值进行类型推测。由于Swift具有类型推测能力,所以代码要比C和Objective-C简洁。

练习题

1. 在日常生活中,哪些信息适合用常量表示?哪些信息适合用变量表示?举例说明,并将这些信息定义为常量或变量。
2. 用变量或常量定义一个圆的半径、圆周率和面积。
3. 检查你给出的定义是否符合变量和常量的命名规则,如何判断?
4. 检查你给出的定义是否符合驼峰拼写法的规则,如何判断?
5. 在你的代码中,将圆的半径赋值为3.0,圆周率赋值为3.1415,面积为圆周率乘以半径的平方。
6. 在你的代码中,圆的半径、圆周率和面积是什么类型?编译器能否判断出来?请说明。
7. 在变量和常量的定义中,将圆的半径、圆周率和面积都设置为Double类型。
8. 向控制台打印输出圆的半径、圆周率和面积的值。注意,打印输出的信息应该具有较好的可读性。
9. 为你的上述代码添加注释,要用到单行注释和多行注释两种方式。
10. 将代码修改为在同一行中定义圆的半径、圆周率和面积。

2.2 整型和浮点型

数字不仅是计算机的底层机器语言,也是软件开发中最重要的内容之一。数字用来表示各种重要的信息,如身高、体重以及身份证号等。数字有两种主要类型,即整型和浮点型。

整型就是用来表示整数的类型,可以是有符号整型,也可以是无符号整型。Swift支持各种位长的整型数,如Int8、Int16、Int32、Int64分别表示位长为8、16、32、64位的有符号整型,而UInt8、UInt16、UInt32、UInt64则分别表示位长为8、16、32、64位的无符号整型。

这里的位长是指计算机里存储的二进制位长。因此，不同位长的整型能存储的最大值各不相同。通过整型的 min 和 max 属性可以方便地获得不同整型的最小值和最大值，如图 2.12 所示。

```
let minValue = Int8.min                    -128

let maxOfInt8 = Int8.max                   127

let maxOfInt16 = Int16.max                 32767

let maxOfInt32 = Int32.max                 2147483647

let maxOfInt64 = Int64.max                 9223372036854775807
```

图 2.12　整型的取值范围

2.2.1　越界赋值

在给不同类型的整数赋值时，要注意其取值范围，不能越界赋值。如图 2.13 所示，编译器给出两条错误警告译成中文："向 UInt8 类型常量 outBoundNum1 赋值为 −16 时发生溢出""向 Int8 类型常量 outBoundNum2 赋值为 160 时发生溢出"。UInt8 是无符号、位长为 8 的整型，其取值范围是[0, 255]，而−16 超出了该范围。Int8 是有符号、位长为 8 的整型，其取值范围是[−128, 127]，而 160 超出了这个范围。因此，这两个越界赋值都是非法的。

图 2.13　越界赋值

2.2.2　越界运算

还有一种情况也会发生越界错误，如图 2.14 所示。常量 outBoundNum1 通过类型转换为 Int8 类型，然后和常量 outBoundNum2 相加，结果为 136，赋值给常量 sum。通过类型推断，编译器知道 sum 的类型为 Int8 类型，因此 sum 的取值范围是[−128, 127]。合法的 Int8 类型整数在运算过程中的运算结果超出了 Int8 的取值范围，从而导致越界错误。越界运算的情况不仅发生在加法运算中，还有可能发生在减法、乘法

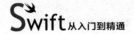

等运算中。

```
let outBoundNum1 : UInt8 = 16
let outBoundNum2 : Int8 = 120
let sum = Int8(outBoundNum1) + outBoundNum2
```
error: Execution was interrupted, reason: EXC_BAD_INSTRUCTION (code=EXC_I386_INVOP, subcode=0x0).
The process has been left at the point where it was interrupted, use "thread return -x" to return to the state before expression evaluation.

图 2.14　越界运算

虽然 Swift 提供了多种整型，但在实际开发中常用到的是整型 Int，它的位长取决于程序运行平台的原生字长。例如，在 64 位的平台上运行 Swift 程序，程序中的 Int 实际就是 Int64。在 32 位的平台上，Int 就是 Int32。在编程过程中使用 Int 不仅灵活，而且还可以提高代码的一致性和可复用性。

2.2.3　整型混合运算

当不同类型的整数进行混合运算时，要先进行类型转换，将其都转换为 Int 类型，然后进行运算，运算结果为 Int 类型，如图 2.15 所示。

```
let num1:Int8 = -128                              -128

let num2:UInt16 = 60000                           60000

let num3: Int32 = -100_0000                       -1000000

let result = Int(num1) + Int(num2) + Int(num3)    -940128
```

图 2.15　不同类型整数的运算

浮点型表示带有小数部分的数的类型。Swift 中提供了两种类型的浮点型 Float 和 Double。Float 为 32 位浮点数类型，一般在对精度要求不高的情况下使用。Double 为 64 位浮点数类型，也叫双精度浮点数类型，应用于描述精度要求高的浮点数。Float 型浮点数的取值范围和精度都要小于 Double 型浮点数，但占用更小的存储空间。现代计算机硬件都能更好地支持 Double 型浮点数运算，因此 Double 型浮点数使用得比较多。

2.2.4　浮点型混合运算

当两种不同类型的浮点数进行混合运算时，也要先进行类型转换，将其都转换为

Double 类型,然后进行运算,运算结果为 Double 类型,如图 2.16 所示。

```
let fNum1: Float = 3.1415                    3.1415
let dNum2: Double = 3.1415926                3.1415926
let sum = Double(fNum1) + dNum2              6.283092596185303
```

图 2.16 不同类型浮点数的运算

练习题

1. 定义常量 maxOfUInt32 和 minOfUInt,并分别对其赋值为 32 位无符号整型的最大值和最小值。

2. 定义一个 Int 型的常量 intNum,赋值为 10000。定义一个 Double 型的常量 doubleNum,赋值为 －33.66。定义一个 Int 型常量 sumInt 和一个 Double 型常量 sumDouble,分别赋值为 intNum 和 doubleNum。

3. 定义一个名为 correctValue 的常量,将其赋值为 200。该常量是什么数据类型? 可以将其声明为 UInt8 和 Int8 吗?

4. 定义两个常量 value1、value2,如图 2.17 所示。将代码输入 playground 中,运行结果如何? 如果发生异常,请分析其原因,并修复错误。

```
let value1 : Int16 = 20000
let value2 : Int8 = 10
let multiResult = value1 * Int16(value2)
```

图 2.17 乘法越界

2.3 布尔型

布尔型是用来描述逻辑上的"真"或"假"的类型。Swift 中,布尔型用关键字 Bool 表示。布尔型只有两个值"真"和"假",分别用关键字 true 和 false 表示。声明布尔型常量或者变量时,既可以显式定义类型,也可以直接赋值。如图 2.18 所示,显式声明变量 daylight 为 Bool 类型时,根据 Swift 4 的语法要求,同时初始化该变量的值为 true。通过直接赋值的方式声明变量 lightSwitch 时,编译器会通过类型推测得到其数据类型。

```
var daylight : Bool = true                    true

var lightSwitch = false                       false
```

图 2.18　类型推测

布尔型主要用于条件语句中，控制程序根据不同的条件执行不同的分支流程。如图 2.19 中，布尔型变量 daylight 用来表示是否为白天。当 daylight 等于 true 时，为白天，执行分支程序，将灯熄灭；当 daylight 等于 false 时，为黑夜，将灯打开。

```
 1  import UIKit
 2
 3  var daylight : Bool = true                 true
 4
 5  var lightSwitch = false                    false
 6
 7  if daylight {
 8      lightSwitch = false                    false
 9      print("It is daytime, so turn  light switch off!")   "It is daytime, so turn  light switch off!\n"
10  } else {
11      lightSwitch = true
12      print("It is night, please turn light switch on!")
13  }
```

It is daytime, so turn light switch off!

图 2.19　布尔型实例

在第 3 章运算符中，还将深入讨论与布尔型相关的逻辑运算符，也叫布尔运算符。

练习题

1. 定义一个布尔型变量 isWeekend，显式声明类型，初始值为 false。根据该变量的值分别打印信息"Today may be Saturday or Sunday.""Today is not Saturday or Sunday."。

2. 定义一个变量 sunnyDay，直接赋值为 true。根据该变量的值分别打印信息"It's sunny.""It's rainy or cloudy."。

2.4　元组型

2.4.1　元组型的定义

元组型是由一个或多个数据类型组成的复合类型。其中，每个数据类型可以是任意类型,并不要求是相同的。

现实生活中有很多数据都适合用元组表示,如矩形(长,宽)可以用元组 rectangle(Int, Int)表示、三维坐标(x, y, z)可以用元组 3D_coordinates(Double, Double, Double)表示、天气(年,月,日,温度)可以用元组 weather(Int, Int, Int, Double)表示。

图 2.20 定义了一个元组类型为(Int, String)的常量 http404Error,用来描述 HTTP 状态码。HTTP 状态码是 Web 服务器返回给网页请求的一个状态值。状态码 404 就表示请求的网页不存在。这里通过元组的方式将状态码及其描述信息绑定在一起,从而增加了代码的可读性。计算机识别元组中的状态码,而用户则阅读该状态码代表的描述信息。

```
import UIKit
let http404Error = (404, "Not Found")
```
(.0 404, .1 "Not Found")

图 2.20 元组型实例

2.4.2 读取元组

那么,如何分别读取元组中不同类型的数据呢?这里介绍 3 种方法。

一种方法是将元组的值赋值给另一个元组。如图 2.21 所示,元组 http404Error 的值赋给了另一个元组(Code, Description)。其中,Code 和 Description 都可以单独读取。如果只想取元组中的部分值,还可以用下画线忽略不需要的部分,例如(_, Information)。

```
let http404Error = (404, "Not Found")
let (Code, Description) = http404Error
print(Code)
print(Description)
```
(.0 404, .1 "Not Found")

"404\n"
"Not Found\n"

图 2.21 读取元组中的值

第二种方法是通过下标访问元组中的特定元素,下标从零开始分别自左向右表示不同的元素,如图 2.22 所示。

```
print(http404Error.0)
print(http404Error.1)
```
"404\n"
"Not Found\n"

图 2.22 打印元组中的值

第三种方法是在定义元组的时候给每个元素命名。读取元组时,就可以通过这些元素的名字获取元素的值,如图 2.23 所示。

```
let http406Error = (Code: 406, Description:
    "Not Acceptable")
print(http406Error.Code)
print(http406Error.Description)
```

```
(Code 406, Description "Not Acceptable")

"406\n"
"Not Acceptable\n"
```

图 2.23　通过名字读取元素的值

元组型在实际应用中是非常有用的一种数据类型,它可作为函数的返回值使用,大大增加了返回信息的可读性,使用起来非常方便。但是,元组并不适合用来创建复杂的数据结构,特别是长期使用的数据类型,这种情况下应该使用结构体或类描述数据结构。

练习题

1. 定义一个元组类型常量 studentInfo,用来存储学生信息,包括学号、姓名和政治面貌。

2. 用书中介绍的 3 种方法分别读取 studentInfo 中的信息,并打印输出。

2.5　可选型

前面介绍的数据类型都必然有一个值与其对应,在其他程序设计语言中也是如此。但是,在 Swift 中还有特殊的数据类型,通常称为可选型 optional。可选型是 Swift 内建安全特性的一个重要体现,这一点在学习完本节内容后就能深刻体会。

可选型用来表示一个变量或常量可能有值,也可能为空的情况。这里用 nil 表示空。需要注意的是,对于整型来说,整数 0 不是 nil,字符串型的空字符串""也不是 nil。nil 可以理解为变量或常量对应的存储空间里没有数据。

例如,学生信息包括姓名、学号、年龄、获奖情况。对于任意一个学生来说,必有姓名、学号和年龄这 3 个信息,但不是每个学生都获得过奖励。姓名、学号和年龄可以分别用字符串型、字符串型和整型描述。而获奖情况比较特殊,有获奖的可以用字符串描述,没有获奖的怎么描述?如图 2.24 所示,在 C 语言里,一般用空字符串""表示。这种解决方法虽然可行,但也容易引起困惑。因为空字符串也是字符串的一个取值,与别的值没有本质差别。这就要求有文档规定空字符串是表示某个特定的情况,否则其他程序员很难理解这个代码。

```
//In objective-C, use String to describe reward
//case 1: Jim has reward
var studentInObjC: (name: String, idNumber: String, age: Int,
    reward: String) = ("Jim", "37060115", 19, "Best Student")
//case 2: Tom has no reward
studentInObjC = ("Tom", "37060115", 20, "")
```

图 2.24　用字符串描述

而 Swift 则通过可选型完美地解决了这个问题。可选型是一种特殊的类型,可以是字符串可选型,也可以是整型可选型。图 2.25 使用字符串可选型描述获奖情况。若学生获得过奖励,则获奖情况的值为字符串,反之则为 nil。

```
//In Swift, use optional to describe reward
//case 1: Jim has reward
var studentInSwift: (name: String, idNumber: String, age: Int,
    reward: String?) = ("Jim", "37060116", 19, "Best Student")
//case 2: Tom has no reward
studentInSwift = ("Tom", "37060115", 20, nil)
```

图 2.25　用字符串可选型描述

2.5.1　定义可选型

声明一个可选型的常量或者变量的格式为

```
let  constantName: Type?=initialValue
var  variableName: Type?=initialValue
```

图 2.26 声明了一个变量 reward,其是字符串可选型,可以赋初值,也可以不赋,默认值为 nil。注意:声明可选型变量时,类型必须显式声明,因为编译器是无法根据初始值推断该变量是否为可选型的。

```
//definition of optional
var reward: String?
reward = "Best Student"
```

nil
"Best Student"

图 2.26　可选型默认值

2.5.2　理解可选型

可选型是一种比较抽象的数据类型,而且是 Swift 中独有的,所以理解起来有一定难度。为了帮助理解，可以将可选型想象成一个盒子。可选型可以表示一个实际的值，也可以表示 nil。这相当于盒子里可以装一个实际的值，也可以空着，但无论装不装实际的值，盒子总是存在的。如图 2.27 所示，学生获奖情况变量是一个字符串可选型，可以看成一个盒子，当有获奖记录时，盒子里就装着获奖信息的字符串 "Best Student"；当没有获奖记录时，盒子就是空的。

图 2.27　可选型的盒子模型

如何打开可选型的盒子，取出其中的数据呢？有 3 种办法：强制拆包、可选型绑定和 nil 聚合运算。

2.5.3　强制拆包

图 2.28 定义了一个字符串可选型的变量 reward，并赋值为字符串 "Best Student"，然后又定义了一个字符串型变量 bonus，并初始化为字符串 "default value"。最后，直接将 reward 赋值给 bonus。运行结果显示中文错误信息"可选型的字符串必须拆包成值类型的字符串"。这说明字符串可选型不能直接赋值给字符串型，要先拆包，变成同类型的数据后才能进行赋值运算。

```
//definition of optional
var reward: String?
reward = "Best Student"

//force unwrapping
var bonus: String = "default value"
bonus = reward
```
Value of optional type 'String?' must be unwrapped to a value of type 'String'　　✕
Coalesce using '??' to provide a default when the optional value contains 'nil'　　Fix
Force-unwrap using '!' to abort execution if the optional value contains 'nil'　　Fix

图 2.28　使用可选型

为了使用可选型的变量或常量,就要用到强制拆包。强制拆包就是在可选型变量或常量后加上"!",可以理解为打开可选型的盒子,并将其中的数据取出来,被取出的数据是具有实际值的数据类型,如字符串型、整型等。如图 2.29 所示,在 reward 变量后加上"!",对其强制拆包,并将结果打印出来。控制台正确输出结果,编译器没有再报错。

```
//definition of optional
var reward: String?                          nil
reward = "Best Student"                      "Best Student"

//force unwrapping
var bonus: String = "default value"          "default value"
bonus = reward!                              "Best Student"
print("\(bonus)")                            "Best Student\n"
```

图 2.29 强制拆包

强制拆包解决了从可选型变量或常量中取出实际值的问题,同时也带来一个新问题。当可选型的盒子里没有任何实际数据(可选型等于 nil)时,对可选型进行强制拆包会导致系统崩溃。如图 2.30 所示,将前面的代码修改一下,去掉 reward 的赋值语句,此时 reward 等于默认值 nil。再次运行程序,系统会报致命中文错误"在拆包一个可选型值时,意外发现该可选型为 nil"。

```
//definition of optional
var reward: String?
//reward = "Best Student"

//force unwrapping
var bonus: String = "default value"
bonus = reward!
pri● error: Execution was interrupted, reason: EXC_BAD_INSTRUCTION (code=EXC_I386_INVOP, subcode=0x0).
```

图 2.30 强制拆包一个等于 nil 的可选型

因此,在强制拆包一个可选型之前,要确认该可选型中存在一个实际值,即该可选型不等于 nil。一个简单直接的方法是用 if 语句判断强制拆包可选型的结果是否等于 nil,如果不相等,则说明可选型的盒子里有实际的值,可以放心使用强制拆包的结果;如果相等,则说明强制拆包的结果为 nil。如图 2.31 所示,在上例的基础上添加 if 语句,再次运行,系统运行正常。

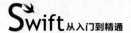

```
//definition of optional
var reward: String?                        nil
//reward = "Best Student"

//force unwrapping
var bonus: String = "default value"        "default value"
//bonus = reward!
print("\(bonus)")                          "default value\n"
if reward != nil {
    bonus = reward!
    print("\(bonus)")
} else {
    print("reward is nil")                 "reward is nil\n"
}
```

图 2.31　用 if 语句判断可选型后再拆包

2.5.4　可选型绑定

可选型绑定就是在 if 或 while 的条件判断语句中，把可选型赋给一个常量，如果可选型含有值，则这条赋值语句的布尔值为 true，同时该常量取得可选型中含有的值；如果可选型等于 nil，则这条赋值语句的布尔值为 false。

格式如下：

```
if let constantName=optionalName {
    some statements
} else {
    some statements
}
```

图 2.32 定义了字符串可选型 reward 和整型可选型 countsOfReward。通过可选型绑定将可选型 reward 和 countsOfReward 分别赋值给同名常量。如果可选型中含有实际的值，则赋值语句的执行结果为 true，将可选型的值赋给常量，然后执行 if 分支。如果可选型为 nil，则赋值语句的执行结果为 false，执行 else 分支。

当有多个可选型要进行绑定时，可以在一个 if 语句中一起绑定。如图 2.33 所示，将上例中的 reward 和 countsOfReward 的可选型绑定合并到一个 if 语句中执行。if 语句后的 3 个语句的执行结果同时为 true 时，执行 if 分支，否则执行 else 分支。

可选型绑定是一种安全的可选型拆包的方法，但需要使用条件判断语句。而强制拆包使用起来简单直接，但却会带来程序崩溃的风险。编写程序时，要根据实际情况确定使用哪种方法进行可选型的取值。在不能确定可选型中是否有值时，应该通过可

```
var reward: String?

//optional binding
if let reward = reward {
    print("The optional's value is \(reward)")
} else {
    print("The optional is nil")
}

var countsOfReward: Int?

countsOfReward = 9

//optional binding
if let countsOfReward = countsOfReward {
    print("Counts of reward is \(countsOfReward)")
} else {
    print("There is no reward record")
}
```

nil

"The optional is nil\n"

nil

9

"Counts of reward is 9\n"

图 2.32　可选型绑定

```
var reward: String?
var countsOfReward: Int?

reward = "Best Student"
countsOfReward = 9

// multiple optionals binding
if let reward = reward, let countsOfReward =
   countsOfReward, countsOfReward > 3 {
    print("reward is \(reward), counts are
        \(countsOfReward)")
} else {
    print("no reward or no counts or counts<=3")
}
```

nil
nil

"Best Student"
9

"reward is Best Student, counts are 9\n"

图 2.33　拆包多个可选型

选型绑定取值。如果能确定可选型中含有值，则可以使用强制拆包的方式。

2.5.5　nil 聚合运算

第 3 种从可选型中取值的方法叫作 nil 聚合运算。这是一种最简便的可选型拆包方式。无论可选型中是否有值，都可以通过 nil 聚合运算拆包。如果可选型中有值，就将其取出来。如果可选型为 nil，就使用默认值替代。

格式为

```
valueFromOptional=optionalType ?? defaultValue
```

这里的"??"为 nil 聚合运算符。当可选型 optionalType 中有值时，将值直接赋给 valueFromOptional；当可选型 optionalType 为 nil 时，将默认值 defaultValue 赋值给 valueFromOptional。

使用 nil 聚合运算符时要注意两点：①默认值的类型必须和可选型拆包后的类型一致；②进行 nil 聚合运算的对象必须为可选型。

如图 2.34 所示，对字符串可选型 reward 和整型可选型 countsOfReward 分别进行 nil 聚合运算。其中，reward 中的值为"Best Student"，所以 valueOfReward 的取值为"Best Student"。而 countsOfReward 为 nil，所以 valueOfCounts 的取值为默认值 0。

```
var reward: String?                              nil
var countsOfReward: Int?                         nil

reward = "Best Student"                          "Best Student"
countsOfReward = nil                             nil

// nil coalescing
var valueOfReward = reward ?? "No Reward"        "Best Student"
var valueOfCounts = countsOfReward ?? 0          0

print(valueOfReward)                             "Best Student\n"
print(valueOfCounts)                             "0\n"
```

图 2.34　nil 聚合运算

实际上，nil 聚合运算是一种可选型绑定的简写方式。如图 2.35 所示，通过可选型绑定实现的代码，与上例通过 nil 聚合运算实现的代码效果一样。但是，nil 聚合运算要简便得多。

```
var reward: String?                              nil
var countsOfReward: Int?                         nil

reward = "Best Student"                          "Best Student"
countsOfReward = 9                               9

var valueOfReward: String = ""                   ""
var valueOfCounts: Int = 3                       3

if let reward = reward, let countsOfReward =
   countsOfReward {
   valueOfReward = reward                        "Best Student"
   valueOfCounts = countsOfReward                9
} else {
   valueOfReward = "No Reward"
   valueOfCounts = 0
}
```

图 2.35　与 nil 聚合运算等价的可选型绑定

练习题

1. 定义一个可选型字符串变量 myHobby，用来表示一个人的兴趣爱好。把你的兴趣爱好，用字符串描述，并赋值给该变量。如果一个人没有任何兴趣爱好，该如何给 myHobby 赋值呢？

2. 定义一个变量 parseStringToInt，将其赋值为 Int("10")，即将一个字符串强制转换为一个整型数，并将其赋值给变量 parseStringToInt。编译器能否推断出变量 parseStringToInt 的类型？是什么类型，为什么是该类型？将另一个字符串 "Picasso" 用同样的方法转换为 Int，并赋值给 parseStringToInt，此时该变量的值为多少？注意，可以选中变量，按住 Option 键，并单击鼠标左键，可查看该变量的类型声明。

3. 定义一个人的名字和兴趣爱好。名字命名为 name，是字符串型，初始化为 "Tommy"。兴趣爱好命名为 hobby，是字符串可选型。用强制拆包的方式取出 hobby 的值，并和 name 的值一起打印。要求：hobby 在不赋值的情况下，系统不报错。

4. 定义一个人的会员资格，命名为 membership，是字符串可选型。向 membership 赋值 "IEEE senior member"。用可选型绑定的方式取出 membership 的值并打印。

5. 在 2、3 题的基础上，向 hobby 赋值 "soccer"。用多重可选型绑定的方式，将 hobby 和 membership 的值一起取出来。当 hobby 和 membership 均不为 nil 时，将它们打印出来。当其中有一个为 nil 时，则打印出错误提示信息。

6. 在 5 题代码的基础上继续定义变量 getHobby 和 getMembership，用 nil 聚合运算的方法分别从可选型变量 hobby 和 membership 中取出其中的值，默认值分别设为 "No hobby" "No membership"。此时，getHobby 和 getMembership 的值是多少？然后，将 hobby 和 membership 赋值为 nil，再重复前面的 nil 聚合运算，观察 getHobby 和 getMembership 中的值是多少，解释原因。

本章知识点

1. 常量和变量的定义以及区别、命名规则、类型声明的格式、打印输出函数 print()、两种添加注释的格式、语句间的分隔、类型推断机制以及驼峰拼写法。

2. 整型和浮点型的位长、整型的 min 和 max 属性、Float 和 Double 的定义、不同类型的混合运算、类型转换、越界赋值以及溢出运算。

3. 布尔型的定义和类型声明,在条件语句中的应用。

4. 元组型的定义和 3 种读取元组的方法。

5. 可选型的 nil 值、定义、盒子模型、强制拆包及其风险、可选型绑定以及 nil 聚合运算。

参考代码

【2.1 节　常量和变量】

参考代码-常量和变量如图 2.36 所示。

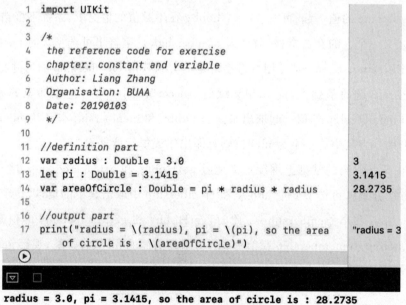

图 2.36　参考代码-常量和变量

【2.2节 整型和浮点型】

第1题的参考代码-整型和浮点型如图2.37所示。

```
let maxOfUInt32 = UInt32.max          4294967295
let minOfUInt = UInt.min              0
```

图2.37 第1题的参考代码-整型和浮点型

第2题的参考代码-整型和浮点型如图2.38所示。

```
let intNum: Int = 10000                            10000
let doubleNum: Double = -33.66                     -33.66
let sumInt = intNum + Int(doubleNum)               9967
let sumDouble = Double(intNum) + doubleNum         9966.34
```

图2.38 第2题的参考代码-整型和浮点型

第3题的参考代码-整型和浮点型如图2.39所示。

```
let correctValue1 = 200                            200
let correctValue2 : UInt8 = 200                    200
let correctValue3 : Int8 = 200  ⓘ Integer literal '200' overflo...
```

图2.39 第3题的参考代码-整型和浮点型

第4题的参考代码-整型和浮点型如图2.40所示。

```
let value1 : Int16 = 20000                         20000
let value2 : Int8 = 10                             10
//let multiResult = value1 * Int16(value2)
let multiResult = Int(value1) * Int(value2)        200000
```

图2.40 第4题的参考代码-整型和浮点型

【2.3节 布尔型】

第1、2题的参考代码-布尔型如图2.41所示。

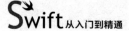

```
var isWeekend : Bool = false                          false

if isWeekend {
    print("Today may be Saturday or Sunday.")
} else {
    print("Today is not Saturday or Sunday.")         "Today is not Saturday or Sunday.\n"
}

var sunnyDay = true                                   true

if sunnyDay {
    print("It's sunny.")                              "It's sunny.\n"
} else {
    print("It's rainy or cloudy.")
}
```

图 2.41　第 1、2 题的参考代码-布尔型

【2.4 节　元组型】

第 1 题的参考代码-元组型如图 2.42 所示。

```
//denifition of tuple
let studentInfo = (60115, "Liang Zhang", true)

//way 1st to read out information from tuple
let (idNum, name, mCCP) = studentInfo
print("Student \(name)'s identification number is \(idNum)")
if mCCP {
    print("\(name) is a member of the Communist Party of China")
} else {
    print("\(name) isn't a member of the Communist Party of China")
}

//way 2nd to read out information from tuple
print("Student \(studentInfo.1)'s identification number is \(studentInfo.0)")
if studentInfo.2 {
    print("\(studentInfo.1) is a member of the Communist Party of China")
} else {
    print("\(studentInfo.1) isn't a member of the Communist Party of China")
}

//way 3rd to read out information from tuple
let anotherStudentInfo = (idNum: 60115, name: "Liang Zhang", mCPC: true)
print("Student \(anotherStudentInfo.name)'s identification number is \(anotherStudentInfo.idNum)")
if anotherStudentInfo.mCPC {
    print("\(anotherStudentInfo.name) is a member of the Communist Party of China")
} else {
    print("\(anotherStudentInfo.name) isn't a member of the Communist Party of China")
}
```

图 2.42　第 1 题的参考代码-元组型

【2.5 节 可选型】

第 1 题的参考代码-可选型如图 2.43 所示。

```
var myHobby: String?                    nil

//I like playing soccer
myHobby = "playing soccer"              "playing soccer"

//I haven't any hobby
myHobby = nil                           nil
```

图 2.43　第 1 题的参考代码-可选型

第 2 题的参考代码-可选型如图 2.44 所示。

```
13   //parse a string into an int
14   var parseStringToInt = Int("10")            10
15
16   parseStringToInt = Int("Picasso")           nil

Declaration
  var parseStringToInt: Int?
Declared In
```

图 2.44　第 2 题的参考代码-可选型

第 3 题的参考代码-可选型如图 2.45 所示。

```
//student's information
var name: String = "Tommy"
var hobby: String?

//force unwrapping
if hobby != nil {
    print("\(name)'s hobby is \(hobby!)")
} else {
    print("\(name) has no hobby")
}
```

图 2.45　第 3 题的参考代码-可选型

第 4 题的参考代码-可选型如图 2.46 所示。

```
var membership: String?
membership = "IEEE senior member"
//optional binding
if let membership = membership {
    print("\(name) is \(membership)")
} else {
    print("\(name) doesn't join any organization!")
}
```

图 2.46　第 4 题的参考代码-可选型

第 5 题的参考代码-可选型如图 2.47 所示。

```
hobby = "soccer"
//multiple optional binding
if let hobby = hobby, let membership = membership {
    print("\(name)'s hobby is \(hobby) and is also
        \(membership)")
} else {
    print("\(name) has no hobby or doesn't join any
        organization!")
}
```

图 2.47　第 5 题的参考代码-可选型

第 6 题的参考代码-可选型如图 2.48 所示。

```
//nil coalescing
var getHobby = hobby ?? "No hobby"
var getMembership = membership ?? "No membership"

hobby = nil
membership = nil

getHobby = hobby ?? "No hobby"
getMembership = membership ?? "No membership"
```

图 2.48　第 6 题的参考代码-可选型

第 3 章 运 算 符

运算符是一种特殊的符号。通过运算符，可以实现对变量或常量的值进行操作，包括值的检查、值的改变及值的合并。本章从最基本的赋值运算符开始，依次介绍算术运算符、关系运算符、逻辑运算符、三元运算符以及区间运算符。

3.1 赋值运算符

赋值运算符"＝"用来初始化或者改变一个变量的值。如图 3.1 所示，变量 str 通过赋值运算符初始化为字符串"Hello,playground"。变量 newStr 通过赋值运算符初始化为变量 str 的值。

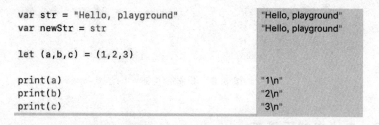

图 3.1 赋值运算符

元组也可以通过赋值运算符，对其中的所有元素一次性赋值。如图 3.1 所示，元组(a,b,c)被初始化赋值为(1,2,3)，相当于 a、b、c 分别被赋值为 1、2、3。

3.2 算术运算符

Swift 中所有的数值类型都支持最基本的四则运算：加（＋）、减（－）、乘（＊）、除（/）。图 3.2 定义了两个整型变量 a 和 b，然后打印出 a 和 b 进行四则运算的结果。字符串类型也可以相加，从而合并成一个新的字符串，通常称之为拼接运算。

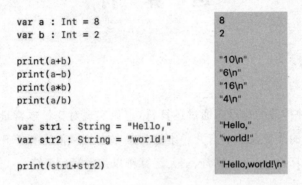

图 3.2 加、减、乘、除运算符

3.2.1 求余运算符

求余运算符"％"也叫取模运算符，其计算两个数相除后的余数。只支持整型求余运算，如图 3.3 所示。

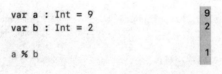

图 3.3 求余运算符

3.2.2 自增自减运算符

Swift 4 中去掉了以前版本中的自增运算符"＋＋"和自减运算符"－－"，取而代之的是"＋＝"和"－＝"。原来的自增/自减运算符是从 C 语言中搬过来，使用起来比较灵活，初学者较难理解。另外，自增运算符"＋＋"主要用于 for 循环语句中。而在 Swift 4 中，for-in 循环结构已经不再需要使用"＋＋"了。

Swift 4 中的自增/自减运算符示例如图 3.4 所示。

```
var a = 3               3

a += 1                  4
let b  = a              4

a -= 1                  3
let c = a               3
```

图 3.4　Swift 4 中的自增/自减运算符示例

3.3　关系运算符

关系运算符就是用来比较两个值之间关系的运算符,运算结果为布尔值,即 true 或者 false。关系运算符包括等于(＝＝)、不等于(！＝)、大于(＞)、小于(＜)、大于或等于(＞＝)、小于或等于(＜＝)。如图 3.5 所示,定义变量 a、b,并赋初值,然后进行各种关系运算,运算结果显示在右侧。关系运算符的运算结果为布尔值,所以关系运算符适用于条件语句。

```
var a = 3, b = 6

a == b                      false
a != b                      true
a > b                       false
a < b                       true
a >= b                      false
a <= b                      true

if a == b {
    print("a is \(a), b is \(b), they are equal.")
} else {
    print("a is \(a), b is \(b), they are not
        equal!")
}
```

"a is 3, b is 6, they are not equal!\n"

图 3.5　关系运算符

3.4 逻辑运算符

逻辑运算符包括与(&&)、或(||)、非(!)，操作数为布尔型。

逻辑非是一元运算符，即操作数为1个，其真值表见表3.1。

表3.1 逻辑非运算真值表

a 的布尔值	!a 的布尔值
true	false
false	true

图3.6定义了电灯开关的状态变量on，初始值为false，即表示灯关着。在if语句中，通过!on判断灯是否关着，如果是，则打印相应信息。

```
var on : Bool = false

if !on {
    print("the light is off")
}
```

false

"the light is off\n"

图3.6 逻辑非运算符

逻辑与是二元运算符，操作数为2个，其真值表为表3.2。

表3.2 逻辑与运算符真值表

a 的布尔值	b 的布尔值	a&&b 的布尔值
true	true	true
true	false	false
false	true	false
false	false	false

图3.7定义了控制电灯开关的函数，需要传入两个参数，分别为布尔变量sunshine（表示是否为白天）和indoors（表示是否家里有人）。表示电灯开关的布尔变量为on，on为true的条件是没有阳光，并且家里有人。将sunshine赋值为false，表示没有阳光。同时，将indoors赋值为true，表示家里有人。调用函数lightControl()，

将变量 sunshine 和 indoors 传入，控制台的输出结果为"turn light on"，即打开电灯。

```
var on : Bool = false                                    false

var sunshine : Bool
var indoors : Bool

func lightControl(sunshine:Bool,indoors:Bool) {
    on = (!sunshine)&&indoors                            true

    if !on {
        print("turn light off")
    } else {
        print("turn light on")                           "turn light on\n"
    }
}
sunshine = false                                         false
indoors = true                                           true
lightControl(sunshine:sunshine, indoors:indoors)
```

图 3.7　电灯开关控制（逻辑与）

逻辑或是二元运算符，操作数是 2 个，真值表见表 3.3。

表 3.3　逻辑或运算符真值表

a 的布尔值	b 的布尔值	a||b 的布尔值
true	true	true
true	false	true
false	true	true
false	false	false

如图 3.8 所示，用逻辑或实现电灯开关的功能，需要对前面的代码做如下修改：将变量名 on 改为 off，当 off 为 true 的时候表示应该关灯。将 on 的逻辑表达式改为 off 的逻辑表达式，它表示有阳光的时候，或者没人在家的时候，将灯关了。程序运行的输出结果完全一致。可以看出，逻辑与和逻辑或的表达式是可以进行逻辑等价转换的。

```
        var off : Bool = false                              false

        var sunshine : Bool
        var indoors : Bool

        func lightControl(sunshine:Bool,indoors:Bool) {

            off = sunshine || (!indoors)                    false

            if off {
                print("turn light off")
            } else {
                print("turn light on")                      "turn light on\n"
            }
        }
        sunshine = false                                    false
        indoors = true                                      true
        lightControl(sunshine:sunshine, indoors:indoors)
```

图 3.8　电灯开关控制（逻辑或）

3.5　三元运算符

前面介绍了多个一元运算符和二元运算符，本节介绍一种三元运算符，即三元条件运算符。

三元条件运算符有 3 个操作数，格式为

question ? answer1 : answer2

其语义为：如果 question 为 true，则返回 answer1；否则返回默认值 answer2。

实际上，三元运算符是一种简写方式，其语义与下面的 if 语句等价。

```
if question: {
    answer1
} else {
    answer2
}
```

图 3.9 定义了 3 个常量 pageHeight、contentHeight、bottomHeight，分别表示页面高度、正文内容高度、底部栏高度，另外还定义了一个布尔型变量 hasHeader，用来表示页面中是否要包含一个头部栏。页面高度由正文内容高度、底部栏高度和头部栏

高度相加得到。如果 hasHeader 为 true 时，头部栏按照 30 算，否则头部栏按照 10 算。这就可以通过三元条件运算符便捷地实现。

```
let contentHeight = 100                        100
let bottomHeight = 20                          20
var hasHeader = true                           true
let pageHeight = contentHeight + bottomHeight + 150
    (hasHeader ? 30 : 10)
```

图 3.9　三元条件运算符

三元条件运算符可以简洁地表示有条件的二选一语义，但是三元条件运算符的缺点是可读性较差。因此，在表达复杂的组合逻辑时，尽量不使用三元条件运算符。

3.6　区间运算符

区间运算符包括闭区间运算符和半闭区间运算符。

闭区间运算符 a...b 表示一个从 a 到 b 的所有值的区间（包括 a 和 b）。闭区间运算符在循环语句中很有用。如图 3.10 所示，在 for-in 循环中，变量 i 取 0～3 的值（包括 0、1、2、3），循环执行 4 次。

半闭区间运算符 a..<b 表示一个从 a 到 b 的所有值的区间（包括 a，但不包括 b）。如图 3.11 所示，在 for-in 循环中，变量 i 取值为 0、1、2，循环执行 3 次。

图 3.10　闭区间运算符　　　　　　　　　图 3.11　for-in 循环

本章知识点

本章介绍了 6 类常用的运算符，分别为赋值运算符、算术运算符（加、减、乘、除、求余、自增、自减）、关系运算符（等于、不等于、大于、小于、大于或等于、小于或等于）、逻辑运算符（与、或、非）、三元运算符以及区间运算符（闭区间运算符和半闭区间运算符）。

第 4 章 字 符 串

字符串是一组有序字符的集合。字符串在移动应用中是一个非常重要的数据类型,在与用户交互的过程中,绝大多数信息都是以字符串的形式呈现的,如个人信息、新闻内容、表格信息等。本章介绍字符串的定义、操作、比较等。

4.1 字符串的定义

4.4.1 字符和字符串

计算机的底层语言是用 0/1 表示的。计算机在理解字符时,是通过将其映射到对应的数字编码实现的。定义字符用关键字 Character 表示。字符类型只能保存一个字符,要表示多个字符,就要字符串类型。字符串是一组有序字符的集合,用 String 关键字表示。图 4.1 定义了 3 个字符常量,分别为字母、数字和符号,另外还定义了这 3 个字符组成的字符串。定义字符类型的常量或变量时,必须显式定义。编译器无法通过类型推断区别该常量或变量是字符类型,还是字符串类型,因为这两种类型的赋值都是用双引号""实现的。

```
let characterAlphabet: Character = "a"          "a"
let characterNumber: Character = "9"            "9"
let characterOperator: Character = "*"          "*"

let str = "a*9"                                 "a*9"
```

图 4.1 字符和字符串的声明

4.1.2 多行文本的赋值

需要输入多行文本时，Swift 提供了一个快捷的方法。如图 4.2 所示，通过三重双引号，可以将多行文本按照原有的换行格式输入并赋值给一个字符串常量 multilinesString。

```
let multilinesString = """
    Do the most simple people,
    the most happy way:
    We often heart will feel tired, just want to too much;
    We always say life trival, is actually I don't understand taste;
    Our business is busy, often is not satisfied;
    We are always competitive, is actually his vanity is too strong.
    Don't compare, heart is cool, life is so simple.
    """
print(multilinesString)
```

图 4.2 多行文本的赋值

4.1.3 字符串初始化

构建长字符串时，通常的做法是先用空字符串作为初始值，然后逐步增加字符串长度。图 4.3 定义了两个字符串变量 emptyStr 和 anotherEmptyStr，其中 emptyStr 的初始化是通过直接赋值空字符串实现的，anotherEmptyStr 则是通过 String()实现，两者的效果一致。

```
var emptyStr = ""
var anotherEmptyStr = String()
```

图 4.3 空字符串

初始化字符串后，可以通过字符串的方法.isEmpty 检查字符串是否为空，如图 4.4 所示。

```
if emptyStr.isEmpty {
    print("This String is empty")
}
```
"This String is empty\n"

图 4.4 空字符串的检查

4.1.4 值类型

字符串类型是值类型。对字符串常量或者变量进行赋值时，是将字符串的值进行了复制，而不是字符串的指针，也就是重新开辟了一片存储空间用来存储常量或变量的字符串，与被复制的字符串分别位于不同的存储空间。因此，用来赋值的字符串变量的值不会随着被赋值的字符串变量值的改变而改变。

如图4.5所示，对字符串变量 str2 赋值为 str1。str2 的值是通过复制 str1 的值得到的，在系统内存中，这两个相同的字符串分别存储在不同的存储空间里。将 str2 的值改为"hello China"后，只是修改了保存 str2 的内存单元，str1 的值不受任何影响。

```
var str1 = "hello world"        "hello world"
var str2 = str1                 "hello world"
str2 = "helllo China"           "helllo China"
print(str1)                     "hello world\n"
```

图 4.5 值类型

练习题

1. 定义两个字符串变量，分别表示大学校名和学院名，并用两种方式赋初始值为空字符串。

2. 在第1题的基础上判断学院名是否为空。如果为空，则打印提示信息；如果不为空，则打印输出学院名。

3. 在第1题的基础上给大学校名赋值一个具体的校名，然后判断其是否为空。如果为空，则打印提示信息；如果不为空，则打印输出校名。

4. 在第3题的基础上定义一个新的字符串变量 myUniversity，并将3题的大学校名赋值给 myUniversity。打印3题中的校名和 myUniversity，然后将 myUniversity 赋值为"Tsinghua University"，再将两个校名打印出来。比较两次打印的结果，并说明原因。

4.2 字符串的操作

下面介绍几个常用的字符串方法。

4.2.1 遍历字符串

在 Swift 中遍历字符串是非常简单的事情,不需要通过字符串的 characters 属性访问。如图 4.6 所示,如果要取出字符串中的字符,通过 for-in 循环直接遍历字符串本身,就可以获得字串符中的每个字符。

```
var welcomeString = "Hello, world!"
for charInString in welcomeString {
    print(charInString)
}
```

"Hello, world!"

(13 times)

图 4.6　遍历字符串

4.2.2 字符数的统计

如图 4.7 所示,要统计字符串中的字符数,直接调用字符串的 count() 方法即可。

```
let countString = welcomeString.count
```

13

图 4.7　字符数的统计

4.2.3 字符串的连接运算

字符串和字符都可以通过加法运算符(＋)进行连接,从而得到一个新的字符串,如图 4.8 所示。

```
var str1 = "Hello"
var str2 = "world"
var character1 = ","
var character2 = "!"

let newString = str1 + character1 + str2 + character2
```

"Hello"
"world"
","
"!"

"Hello,world!"

图 4.8　字符串连接

4.2.4 插入字符串

如果要向一个字符串中插入一个常量或变量,可以采用格式:\(constantName or variableName)。如图 4.9 所示,整型常量 insertedNum 的值被插入到字符串中。

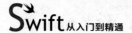

```
let insertedNum = 888
let insertedString = "The number \(insertedNum) is
    inserted into this string!"
```

```
888
"The number 888 is inserted into this string!"
```

图 4.9 插入字符串

4.2.5 字符串的大小写

另外,还可以通过字符串的方法 uppercased() 和 lowercased() 分别取得字符串的大写和小写版本,如图 4.10 所示。

```
let theString = "It's my world!"
print("the uppercaseString is \(String(describing:
    theString.uppercased))")
print("the lowercaseString is \(String(describing:
    theString.lowercased))")
```

```
"It's my world!"
"the uppercaseString is (Function)\n"
"the lowercaseString is (Function)\n"
```

图 4.10 字符串的大小写

4.2.6 字符串索引

字符串中的字符可以通过字符串索引获取。在其他编程语言中,字符串的索引一般都为整型,而 Swift 的字符串索引则为一种特殊的类型 String.Index。这使得字符串索引的相关操作复杂一些。

要获取字符串中某个位置的字符时,首先要获取字符串中特殊位置的索引,如起始位置或结束位置。图 4.11 定义了一个字符串常量 courseTitle,用来保存课程名,并赋值为"Swift"。然后,通过字符串的方法 startIndex() 获得字符串的起始位置索引,并赋值给常量 firstIndex,它的类型为 String.Index。最后,通过 firstIndex 获取字符串中起始位置的字符"S"。

```
let courseTitle = "Swift"
let firstIndex = courseTitle.startIndex
let firstCharacter = courseTitle[firstIndex]
```

```
"Swift"
String.Index
"S"
```

图 4.11 字符串起始位置索引

这里要特别注意,字符串中最后一个字符的位置在结束位置前,如果根据结束位置索引获取字符,会导致严重错误。图 4.12 定义了一个常量 lastIndex,赋值为字符串的结束位置索引。然后,根据该索引获取对应位置的字符,结果导致系统出错。

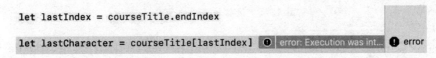

图 4.12　字符串结束位置索引

取出字符串结束位置字符的正确方式是根据字符串结束位置前的索引,获取字符串的最后一个字符,如图 4.13 所示。

```
//let lastIndex = courseTitle.endIndex
let lastIndex = courseTitle.index(before: courseTitle.endIndex)     String.Index
let lastCharacter = courseTitle[lastIndex]                          "t"
```

图 4.13　获取结束位置字符的正确方式

在起始位置或结束位置的索引基础上,可以通过指定偏移量获取其他位置的字符。如图 4.14 所示,定义常量 middleIndex 存储字符串的中间位置的索引。这里以字符串起始位置为基准,向后偏移量为 2,即字符串的第 3 个字符的位置,也就是中间位置,然后再根据 middleIndex 索引获取中间位置的字符"i"。

```
let middleIndex = courseTitle.index(courseTitle.startIndex,      String.Index
    offsetBy: 2)
let middleCharacter = courseTitle[middleIndex]                   "i"
```

图 4.14　获取字符串中任意位置的字符

4.2.7　字符串的子串

字符串的子串就是从字符串中截取一部分,形成一个新的字符串。例如,姓名是一个人基本信息中常见的字符串,它的子串"姓"和"名"就是有用的两个子串。如何获取字符串中的特定子串呢?图 4.15 定义了表示学生姓名的字符串常量 studentName,并赋了初始值。要获取学生的姓和名,可以先得到字符串中空格的字符串索引。表示名的子串就是字符串中从起始位置到空格位置的子串。注意,起始位置属于子串的一部分,而空格则不属于子串,因此这里使用符号"<"表示小于该索引值。

这里使用了区间标识 [startPosition, endPosition],表示从 startPosition 到

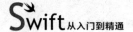

```
let studentName = "Liang Zhang"
let spaceIndex = studentName.firstIndex(of: " ")!
let firstName = studentName[studentName.startIndex..<spaceIndex]
```

图 4.15　获取姓名中的子串

endPosition 之间的所有元素，包括 startPosition 和 endPosition。其中，startPosition 为区间的起始位置索引，endPosition 为区间的结束位置索引。

如果 startPosition 为字符串的起始位置，或者 endPosition 为字符串的结束位置，则可采用半开区间写法，默认表示从起始位置到特定位置，或者从特定位置到结束位置。如图 4.16 所示，学生的名用常量 firstName2 表示，姓用常量 surname 表示。这两个子串分别是从起始位置到空格和从空格到结束的位置，均采用半开区间表示子串的范围。

```
let firstName2 = studentName[..<spaceIndex]
let surname = studentName[studentName.index(after: spaceIndex)...]
```

图 4.16　半开区间表示子串的范围

检查子串的类型声明发现，子串并不是字符串类型，而是一种特殊的类型 String.SubSequence。当要将子串作为字符串使用时，需要将其强制转换为字符串，如图 4.17 所示。子串之所以不同于字符串，是因为子串与其对应的字符串指向了同一块物理空间，这个空间存储整个字符串的内容，当然也包括子串的内容。当子串被强制转换为字符串后，就重新开辟一块新的物理空间，用来存储子串的内容。

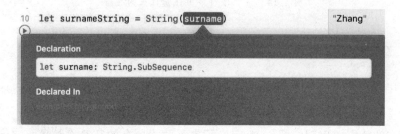

图 4.17　字符串子串的类型

4.2.8　字符串的比较

可以从 3 个维度比较字符串，即字符串相等、字符串前缀相等和字符串后缀相等。

如果两个字符串中的字符以相同顺序出现,并且完全相同,则两个字符串相等,如图 4.18 所示。

```
var compareStr1 = "It is compare string."
var compareStr2 = "It is compare string."
if compareStr1 == compareStr2 {
    print("they are equal!")
}
```

"It is compare string."
"It is compare string."

"they are equal!\n"

图 4.18 字符串相等

要判断字符串是否含有特定的字符串前缀,可以使用方法 hasPrefix()。通过该方法,可以便捷地判断一个字符串是否包含特定的字符串前缀,具体格式为

theString.hasPrefix(thePrefix)

图 4.19 定义了字符串常量 bookInfo1。通过 hasPrefix() 方法判断该字符串是否包含特定的前缀"History book"。

```
let bookInfo1 = "History book: world history"

if bookInfo1.hasPrefix("History book") {
    print("book1 is a history book.")
}
```

"History book: world history"

"book1 is a history book.\n"

图 4.19 判断字符串的特定前缀

要判断字符串中是否含有特定的字符串后缀,可以使用方法 hasSuffix(),具体格式为

theString.hasSuffix(theSuffix)

图 4.20 定义了字符串常量 bookInfo2,并检索了该字符串是否包含特定的后缀"history"。

```
let bookInfo2 = "History book: China history"

if bookInfo2.hasSuffix("history") {
    print("book2 is a history book.")
}
```

"History book: China history"

"book2 is a history book.\n"

图 4.20 判断字符串的特定后缀

练习题

1. 定义一个字符串变量 myHobbies,并将其赋值为你的兴趣爱好,至少为 3 个。遍历 myHobbies 字符串中的每个字符并打印,查看循环执行的次数。

2. 在第 1 题的基础上调用字符串的字符数统计方法,查看字符数是否和 1 题的循环执行次数一致,为什么?

3. 定义两个字符串常量 str1 和 str2,分别赋值为 "My hobbies" "is"。定义一个字符串变量 info,赋值为 3 个字符串 str1、str2、myHobbies 连接的结果。打印字符串 info。

4. 在第 3 题的基础上向字符串 info 中的单词之间插入空格,并重新打印输出。

5. 打印输出字符串 info 的大写和小写版本。

6. 将字符串变量 myHobbies 中的第一个字母和最后一个字母取出并打印。myHobbies 中有 3 个兴趣,将第二个兴趣的首字母,通过字符串索引打印出来,要求分别从字符串的起始位置和结束位置向第二个兴趣的首字母检索。

7. 运用字符串子串中的相关知识将 myHobbies 中的 3 个兴趣分别保存到 3 个新的字符串中,并打印输出,注意每个兴趣中都不要含有多余的空格。

8. 比较第 7 题中的第一、二个兴趣的字符串是否相等,并打印输出结果。检索第一个兴趣中是否含有特定字符串前缀,检索第二个兴趣中是否含有特定字符串后缀,并打印输出结果。

本章知识点

1. 字符和字符串的定义、多行文本的赋值方法、字符串的初始化及空字符串检测、字符串的值类型特点。

2. 字符串的常用操作包括遍历字符串、字符数的统计、字符串的连接运算、插入字符串、字符串的大小写、字符串索引、字符串的子串、字符串的比较(字符串相等、字符串前缀相等、字符串后缀相等)。

参考代码

【4.1 节 字符串的定义】

第 1 题的参考代码-字符串的定义如图 4.21 所示。

```
var schoolName = ""
var universityName = String()
```

图 4.21　第 1 题的参考代码-字符串的定义

第 2 题的参考代码-字符串的定义如图 4.22 所示。

```
if schoolName.isEmpty {
    print("School Name is empty!")
} else {
    print("My School is \(schoolName)")
}
```

图 4.22　第 2 题的参考代码-字符串的定义

第 3 题的参考代码-字符串的定义如图 4.23 所示。

```
universityName = "Beijing University of Aeronautic and
    Astronautic"
if universityName.isEmpty {
    print("University Name is empty!")
} else {
    print("My University is \(universityName)")
}
```

图 4.23　第 3 题的参考代码-字符串的定义

第 4 题的参考代码-字符串的定义如图 4.24 所示。

```
var myUniversity = ""
myUniversity = universityName
print("My university is \(myUniversity), University Name
    is \(universityName)")

myUniversity = "Tsinghua University"
print("My university is \(myUniversity), University Name
    is \(universityName)")
```

图 4.24　第 4 题的参考代码-字符串的定义

【4.2 节　字符串的操作】

第 1 题的参考代码-字符串的操作如图 4.25 所示。

```
var myHobbies = "Music Basketball Philosophy"

for char in myHobbies {
    print(char)
}
```

图 4.25　第 1 题的参考代码-字符串的操作

第2题的参考代码-字符串的操作如图4.26所示。

```
print("The number of characters in myHobbies is \(myHobbies.count)")
```

图4.26　第2题的参考代码-字符串的操作

第3题的参考代码-字符串的操作如图4.27所示。

```
let str1 = "My hobbies", str2 = "is"
var info = str1 + str2 + myHobbies
print(info)
```

图4.27　第3题的参考代码-字符串的操作

第4题的参考代码-字符串的操作如图4.28所示。

```
let space = " "
info = str1 + space + str2 + space + myHobbies
print(info)
```

图4.28　第4题的参考代码-字符串的操作

第5题的参考代码-字符串的操作如图4.29所示。

```
print(info.uppercased())
print(info.lowercased())
```

图4.29　第5题的参考代码-字符串的操作

第6题的参考代码-字符串的操作如图4.30所示。

```
let firstIndex = myHobbies.startIndex
let firstChar = myHobbies[firstIndex]

let lastIndex = myHobbies.index(before: myHobbies.endIndex)
let lastChar = myHobbies[lastIndex]

var middleIndex = myHobbies.index(firstIndex, offsetBy: 6)
var middleChar = myHobbies[middleIndex]
middleIndex = myHobbies.index(lastIndex, offsetBy: -20)
middleChar = myHobbies[middleIndex]
```

图4.30　第6题的参考代码-字符串的操作

第7题的参考代码-字符串的操作如图4.31所示。

```swift
var spaceIndex = myHobbies.firstIndex(of: " ")!
let hobby1st = myHobbies[firstIndex..<spaceIndex]

firstIndex = myHobbies.index(after: spaceIndex)
let restOfMyHobbies = myHobbies[firstIndex...]
spaceIndex = restOfMyHobbies.firstIndex(of: " ")!

let hobby2nd = restOfMyHobbies[restOfMyHobbies.startIndex..<spaceIndex]
firstIndex = restOfMyHobbies.index(after: spaceIndex)
let hobby3rd = restOfMyHobbies[firstIndex...]

print("The first hobby is \(hobby1st)")
print("The second hobby is \(hobby2nd)")
print("The third hobby is \(hobby3rd)")
```

图4.31　第7题的参考代码-字符串的操作

第8题的参考代码-字符串的操作如图4.32所示。

```swift
//compare hobby
if hobby1st == hobby2nd {
    print("The two hobbies are equal.")
}

if hobby1st.hasPrefix("musi") {
    print("The first hobby: \(hobby1st) has prefix: musi")
} else {
    print("The first hobby: \(hobby1st) hasn't prefix: musi")
}

if hobby2nd.hasSuffix("sophy") {
    print("The last hobby: \(hobby2nd) has suffix: sophy")
} else {
    print("The last hobby: \(hobby2nd) hasn't suffix: sophy")
}
```

图4.32　第8题的参考代码-字符串的操作

第 5 章 集 合

集合就是用来存储一组数据的容器。Swift 中有 3 种常用的集合类型：数组、集合和字典。

数组是一种按顺序存储相同类型数据的集合，相同的值可以在数组中的不同位置重复出现。集合和字典类型也是存储了相同类型数据的集合，但是数据之间是无序的。集合不允许值重复出现。字典中的值可以重复出现，但是每个值都有唯一的键值与其对应。本章将对这 3 种非常重要的集合类型做详细介绍。

5.1 数组

当要按照一定的次序存储一组数据，并通过数据所在位置的序号检索时，就要采用数组实现。数组是类型安全的，因此数组中包含的数据类型必须是显式声明的。

数组的声明格式为 Array<DataType> 或者 [DataType]。后一种声明方式比较简洁，用得较多。

5.1.1 数组的初始化

数组有多种初始化方法，这里介绍 3 种。

第一种初始化方法是先将数组声明为一个空数组，然后再逐步往数组中增加元素。如图 5.1 所示，首先定义一个字符串型的空数组变量 animalArray，然后调用数组类型的方法 isEmpty() 判断数组是否为空。如果为空，则开始以此向数组中添加元素。

添加元素用到数组类型的方法 append()。append()方法用来向数组的末端添加一个元素。需要注意的是，添加的元素必须和数组声明的数据类型一致，否则会导致报错。另外，append()方法每次都是向数组的末端添加元素，因此数组中的元素是按照添加的次序顺序排列的。

```
var animalArray = [String]()                    []
if animalArray.isEmpty {
    print("animalArray is empty!")              "animalArray is empty!\n"
}
animalArray.append("tiger")                     ["tiger"]
animalArray.append("lion")                      ["tiger", "lion"]
```

图 5.1　isEmpty()方法和 append()方法

第二种数组初始化的方法是在声明数组的时候直接赋值为一个实际的数组。图 5.2定义了一个整型数组变量 oneBitNumberArray，并在声明语句中直接将一个含有 10 个整型数的数组赋值给它。另外，还用同样的方式定义了两个字符串型数组 botanyArray1 和 botanyArray2。最后，调用数组类型的 count()方法统计并打印数组 botanyArray1 的长度（即数组所含元素的个数）。

```
var oneBitNumberArray : Array<Int> =            [0, 1, 2, 3, 4, 5, 6, 7, 8, 9]
    [0,1,2,3,4,5,6,7,8,9]
var botanyArray1 : [String] =                   ["rosemary", "parsley", "sage", "thyme"]
    ["rosemary","parsley","sage","thyme"]
var botanyArray2 =                              ["rosemary", "parsley", "sage", "thyme"]
    ["rosemary","parsley","sage","thyme"]
print("There are \(botanyArray1.count) kinds    "There are 4 kinds of botany.\n"
    of botany.")
```

图 5.2　数组初始化

第 3 种数组初始化的方法叫作快捷初始化方法，即一次性将数组的所有元素初始化为同一个值。图 5.3 定义了两个整型数组变量 twoBitNumberArray 和 threeBitNumberArray，在声明语句中同时定义了数组的长度（count）分别为 6 和 3，以及每个元素的初值（repeating）分别为 0 和 11。

```
var twoBitNumberArray = [Int](repeating: 0,     [0, 0, 0, 0, 0, 0]
    count: 6)
var threeBitNumberArray = [Int](repeating:      [11, 11, 11]
    11, count: 3)
```

图 5.3　数组的相加

5.1.2 数组的相加和累加运算

数组的相加和累加运算分别使用运算符"＋"和"＋＝"，从而由已知数组快速得到新数组。图5.4定义了一个数组变量theAddedNumberArray，该数组由两个整型数组 twoBitNumberArray 和 threeBitNumberArray 相加而得。字符串数组animalArray中含有两个元素，然后通过两次累加分别将含有一个元素和两个元素的数组添加到animalArray数组中。这里需要注意的是，相加或累加的数组必须是相同数据类型的。

```
var theAddedNumberArray = twoBitNumberArray +      [0, 0, 0, 0, 0, 0, 11, 11, 11]
    threeBitNumberArray

animalArray += ["hawk"]                            ["tiger", "lion", "hawk"]
animalArray += ["sheep","horse"]                   ["tiger", "lion", "hawk", "sheep", "horse"]
```

图5.4 数组的累加

5.1.3 数组的下标操作

数组中的元素是有序排列的，通过下标可以找到特定位置的元素，获取该元素的值或者对其进行修改。注意，数组中第一个元素的下标序号是0，而不是1。如图5.5所示，要取出字符串数组animalArray中第一个元素的值，只通过animalArray[0]即可。要修改数组中第二个元素的值，直接对animalArray[1]进行赋值即可。另外，还可以通过下标批量修改数组元素的值。

```
var theFirstAnimal = animalArray[0]                "tiger"
animalArray[1] = "hen"                             "hen"
animalArray[2...4] =                               ["goat", "rat", "cock", "rabbit"]
    ["goat","rat","cock","rabbit"]
print(animalArray)                                 ["tiger", "hen", "goat", "rat", "cock", "rabbit"]\n"
```

图5.5 数组的下标操作

5.1.4 插入与删除元素

数组的插入和删除操作分别用insert()和remove()方法。如图5.6所示，向数组animalArray中下标为3的位置插入一个新元素字符串"snake"。如果删除数组中第一个元素"tiger"或者最后一个元素"rabbit",可以直接调用方法 removeFirst()和 removeLast()。如果要删除特定位置的元素,可以调用方法 remove(at: Int),并给出该位置的具体参数。

```
print(animalArray)                          "["tiger", "hen", "goat", "rat", "cock", "rabbit"]\n"
animalArray.insert("snake", at: 3)          ["tiger", "hen", "goat", "snake", "rat", "cock", "rabbit"]

animalArray.removeFirst()                   "tiger"
print(animalArray)                          "["hen", "goat", "snake", "rat", "cock", "rabbit"]\n"

animalArray.removeLast()                    "rabbit"
print(animalArray)                          "["hen", "goat", "snake", "rat", "cock"]\n"

animalArray.remove(at: 2)                   "snake"
print(animalArray)                          "["hen", "goat", "rat", "cock"]\n"
```

图 5.6　数组的插入和删除

5.1.5　数组的遍历

数组的遍历一般采用 for-in 语句。如图 5.7 所示,第一个 for-in 循环对数组 animalArray 中的元素进行遍历并打印。第二个 for-in 循环以元组的方式对数组中的下标和相应元素遍历并打印。

```
for animal in animalArray {
    print(animal)                                           (4 times)
}

for (index,animal) in
    animalArray.enumerated() {
    print("No.\(index) animal is \(animal)")                (4 times)
}
```

图 5.7　数组的遍历

5.1.6　数组片段

通过数组的下标操作可以获取数组中的某一个元素,如果需要获取数组的一部分或者多个元素,就要用到获取子数组的方法。

获取数组 Array 的片段 arraySlice 的格式为

arraySlice=Array[startIndex...endIndex]

其中 startIndex 为子数组起始元素的索引，endIndex 为结束元素的索引。注意，startIndex 的最小值为 0，即 Array 的第一个元素的索引，endIndex 的最大值为 Array.count−1，即 Array 的最后一个元素的索引。

数组片段是由原数组中部分连续的元素组成的子集。数组片段是原数组的一部分，共用相同的存储空间。换言之，如果数组片段中元素的值发生变化，那么原数组中相应元素的值也会同步改变。

如图 5.8 所示，数组 animalArray 中含有 4 个元素。常量 myZoo 赋值为数组 animalArray 的一个片段，包含下标 1~2 的两个元素。但是，myZoo 不是一个新的数组，myZoo[0]元素并不存在。myZoo 只是原数组的一部分，它的类型是 ArraySlice。myZoo 中元素的下标和数组 animalArray 中相应元素的下标一致。

```
print(animalArray)                    ["hen", "goat", "rat", "cock"]\n

let myZoo = animalArray[1...2]        ["goat", "rat"]
print(myZoo)                          ["goat", "rat"]\n
myZoo[1]                              "goat"
myZoo[2]                              "rat"
//myZoo[0]
```

图 5.8　数组片段

如果要根据数组片段生成一个新的数组，可以将数组片段类型转换为数组。如图 5.9 所示，将数组片段 animalArray[1...2]类型转换为数组后，数组 newZoo 中的元素就从下标 0 开始排列了。

```
let newZoo = Array(animalArray[1...2])    ["goat", "rat"]
print(newZoo)                             ["goat", "rat"]\n
newZoo[0]                                 "goat"
newZoo[1]                                 "rat"
```

图 5.9　由数组片段生成新数组

5.1.7　元素交换位置

在数组中，如果要交换两个元素的位置，可以使用方法 swapAt(_:_:)，参数表中提供需要的元素位置。如图 5.10 所示，对数组 animalArray 中下标为 2 和 3 的元素交换位置。

```
print(animalArray)
animalArray.swapAt(2, 3)
```

`"["hen", "goat", "rat", "cock"]\n"`
`["hen", "goat", "cock", "rat"]`

图 5.10　数组元素交换位置

5.1.8　数组排序

如果要对数组中的所有元素进行排序，可以使用方法 sort()。如图 5.11 所示，对数组 animalArray 中的所有元素按照字母先后顺序排序。

```
print(animalArray)
animalArray.sort()
```

`"["hen", "goat", "cock", "rat"]\n"`
`["cock", "goat", "hen", "rat"]`

图 5.11　数组元素排序

5.1.9　检索特定元素

如果要检索数组或数组片段中是否含有特定的元素，可以使用方法 contains(_:)。该方法根据检索结果返回布尔值。图 5.12 分别检索了数组片段 animalArray[1...3] 和数组 animalArray 中是否含有字符串"hen"，检索结果为 true 或 false。

```
animalArray[1...3].contains("hen")
animalArray.contains("hen")
```

`false`
`true`

图 5.12　检索数组中是否含有特定元素

练习题

1. 定义 3 个字符串数组，分别用来表示已经选修的专业必修课、专业选修课以及通识课。分别用 3 种方法对这 3 个数组进行初始化，其中专业必修课包含数据结构、计算机组成、计算机网络；专业选修课包含 iOS 应用开发实践、Swift 程序设计、人工智能；通识课包含音乐、美术、文学。

2. 定义一个字符串数组，用来表示所有的选修课程，由专业必修课、专业选修课和通识课组成。

3. 在所有选修课程数组中，将专业必修课"数据结构"修改为"离散数学"，增加一门新的专业选修课"移动应用开发实践"，删除通识课"美术"。

4. 将所有选修课程数组中的元素及其下标依次打印输出。

5. 在第 4 题所有选修课程数组中，生成下标 3～6 元素的数组片段，由该数组片段生成一个新的数组。该数组片段和这个新数组是否完全相同？请通过代码说明两者的相同和不同之处。

6. 对第 5 题中新数组的第一个元素和最后一个元素交换位置，并打印输出。

7. 对新数组中的所有元素进行全排序，并打印输出。

8. 检索新数组中是否含有字符串"Swift"和"Swift Programming"，并打印输出检索结果。

5.2 集合

集合是由一组相同数据类型的元素组成的，每个元素的值都是唯一的。集合中的元素是无序的。

集合类型的声明格式为 Set<DataType>。

5.2.1 集合的初始化

集合类型的初始化既可以从创建一个空的集合开始，也可以通过直接赋值的方法。第一种方法需要显式地指出集合中的元素类型。第二种方法可以通过被赋的初始值推断集合中的元素类型。如图 5.13 所示，通过两种初始化方法定义了集合类型变量 weatherOfSanya 和 weatherOfBj。

```
var weatherOfSanya = Set<String>()
weatherOfSanya = ["rainy","sunny","stormy"]

var weatherOfBj : Set = ["dry","windy","frogy"]
```

图 5.13 集合的初始化

5.2.2 集合的为空判断和插入元素

集合类型和数组类型一样，也提供了 isEmpty()和 insert()方法，可以便捷地操作集合。如图 5.14 所示，通过方法 isEmpty()判断集合 weatherOfSanya 是否为空，并打印提示信息。然后，通过方法 insert(_:)向集合中添加一个元素"cloudy"。

```
if weatherOfSanya.isEmpty {
    print("The set of weather is empty!")
} else {
    print("There are \(weatherOfSanya.count) kinds weather!")
}

weatherOfSanya.insert("cloudy")
```

图 5.14　集合的 isEmpty() 和 insert() 方法

5.2.3　删除元素

要删除集合中的一个元素，可以使用方法 remove(_:)。在 remove() 方法的参数表中要提供被移除元素的值，如果集合中有这个元素，则删除后返回这个元素的值；如果集合中不存在该元素，则返回一个 nil。要删除集合中的所有元素，可以直接调用方法 removeAll()。如图 5.15 所示，集合变量 weatherOfSanya 中含有元素"stormy"，删除该元素后，返回值为"stormy"。然后删除元素"dry"时，由于集合 weatherOfSanya 中不含有此元素，因此返回值为 nil。

```
weatherOfSanya.remove("stormy")
weatherOfSanya.remove("dry")
```

图 5.15　集合中删除元素

5.2.4　检索特定元素

要检查集合中是否包含某个特定的元素，可以直接调用方法 contains(:_)，如图 5.16 所示。

```
if weatherOfSanya.contains("sunny") {
    print("Sanya is sunny sometimes.")
}
```

图 5.16　集合中检索特定元素

5.2.5　遍历集合

通过 for-in 循环可以对集合进行遍历，如图 5.17 所示。

```
for weather in weatherOfSanya {
    print("\(weather)")
}
```

图 5.17　集合的遍历

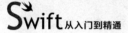

5.2.6 集合排序

集合中的元素是乱序排列的。如果要有序地输出集合元素,可以通过 sorted()函数对集合中的元素按照值先排序,然后再输出。图 5.18 打印了集合变量 weatherOfSanya 排序后的结果。

```
print("the result of unsorted : ")
for weather in weatherOfSanya {
    print("\(weather)")
}
```

图 5.18 集合的排序

5.2.7 集合间的运算

Swift 提供各种方法支持两个集合之间的操作,包括 intersect()、union()、subtract()。其中,intersect()方法计算两个集合的交集,结果为一个新的集合;union()方法计算两个集合的并集,结果为一个新的集合;subtract()方法计算一个集合中不属于另一个集合的部分,结果为该部分元素形成的新集合。图 5.19 对两个集合变量 weatherOfSanya 和 weatherOfBj 分别进行交集、并集和相减运算。

```
var weatherOfSanya = Set<String>()
weatherOfSanya = ["rainy","sunny","stormy"]

var weatherOfBj : Set = ["dry","windy","frogy","sunny"]

weatherOfBj.intersection(weatherOfSanya)
weatherOfBj.union(weatherOfSanya)
weatherOfBj.subtracting(weatherOfSanya)
weatherOfBj.subtract(weatherOfSanya)
```

图 5.19 集合的实例

练习题

1. 定义一个空的字符串型集合 gradeOfTheory,用来表示理论课成绩的分档集合,然后判断 gradeOfTheory 是否为空。如果不为空,则打印提示信息;如果为空,则依次向集合中添加元素 Fail、Pass、Common、Good、Excellent。

2. 定义一个字符串型集合 gradeOfExperiment,用来表示实验课成绩的分档集合，并直接将 gradeOfTheory 赋值给 gradeOfExperiment。然后，将其中的元素 Common、Good、Excellent 从集合中删除。最后，判断集合 gradeOfExperiment 中是否含有元素 Fail 和 Pass，并打印出提示信息。

3. 分别遍历集合 gradeOfTheory 和 gradeOfExperiment，并打印出其中的元素。

4. 对集合 gradeOfTheory 和 gradeOfExperiment 进行排序后，重新打印。

5. 对集合 gradeOfTheory 和 gradeOfExperiment 进行交集、并集、相减运算，并将结果打印出来。集合的相减运算相关的方法有 subtracting() 和 subtract()，请说明两者的差别。

5.3 字典

字典通过键值对（key-value pair）的形式存储数据，每个值（value）都有唯一的键（key）与其对应。字典中数据的组织是无序的。对字典中元素的操作，一般都是基于 key 实现的。

声明字典类型的格式为 Dictionary<KeyType，ValueType>。其中，KeyType 为 key 的类型，ValueType 为 value 的类型。

声明字典类型的简化格式为[KeyType：ValueType]。

5.3.1 字典的初始化

字典的初始化从声明并创建一个空字典开始。如图 5.20 所示，分别通过两种声明格式定义并初始化了两个字典类型变量。其中，字典变量 ascIIDictChar 的键为整型，值为字符型；字典变量 ascIIDictNum 的键为整型，值为整型。

```
var ascIIDictChar = Dictionary<Int, Character>()
var ascIIDictNum = [Int:Int]()
```

图 5.20　字典的声明

声明字典变量的时候，如果知道字典中的元素，可以直接初始化为具体的值。如图 5.21 所示，在定义两个字典类型变量时直接赋初始值。其中，ascIIDictChar 字典中包含 6 个元素，键为十进制的 ASCII 码值，值为码值对应的字符。ascIIDictNum 字典中也包含 6 个元素，键为十进制的 ASCII 码值，值为码值对应的整型数字。

```
var ascIIDictChar = [97:"a",98:"b",99:"c",100:"d",101:"e",102:"f"]
var ascIIDictNum = [32:0,33:1,34:2,35:3,36:4,37:5,38:6]
ascIIDictNum.reserveCapacity(10)
```

图 5.21　字典的初始化

如果事先能确定字典的长度,在字典变量初始化后可以设置它的容量。如图 5.21 所示,规定了字典变量 ascIIDictNum 为 10。通过设置字典变量的容量,编译器可以分配相应的资源给它,从而起到优化性能的作用。

5.3.2　两种更新字典元素的方法

对字典元素进行操作,一般有两种方法:一种是利用下标语法;另一种是利用字典类型自带的方法和属性。这两种方法的使用和数组中下标和方法的使用类似。

如图 5.22 所示,先用下标语法创建一个新的键值对,键为 103,值为 "g",然后修改已经存在的键值对 [97:"a"] 为 [97:"A"]。

```
ascIIDictChar[103] = "g"
print(ascIIDictChar)
ascIIDictChar[97] = "A"
```

图 5.22　字典的键值对

如图 5.23 所示,使用字典类型的方法 updateValue() 将键为 97 对应的值更新为 "a",注意观察字典变量 ascIIDictChar 前后打印值的变化。updateValue() 方法是对特定的键值进行修改。如果存在这个键,则修改成功并返回该键原来对应的值;如果不存在该键,则返回 false。

```
print(ascIIDictChar)                                   ["97": "A", 99: "c", 103: "g", 98: "b", 102: "f", 100: "d", 101: "e"]\n
if let originValue = ascIIDictChar.updateValue("a",
    forKey: 97) {                                      "The origin value is A\n"
    print("The origin value is \(originValue)")
}                                                      ["97": "a", 99: "c", 103: "g", 98: "b", 102: "f", 100: "d", 101: "e"]\n
print(ascIIDictChar)
```

图 5.23　字典的 updateValue() 方法

5.3.3　两种删除字典元素的方法

删除字典中的一个元素也有两种方法:第一种方法是给特定键对应的值赋值为 nil,相当于删除了这个键值对;第二种方法是用字典类型的方法 removeValue()。

第 5 章 集合　Chapter 5

图 5.24 给出了两种删除键值对的实例。removeValue()方法是删除特定的键值。如果存在这个键，则删除成功并返回该键原来对应的值；如果不存在该键，则返回 false。

```
print(ascIIDictChar)
ascIIDictChar[97] = nil
print(ascIIDictChar)

if let removedValue =
    ascIIDictChar
    .removeValue(forKey: 98){
    print("Value
        \(removedValue) is
        removed.")
}
```

```
"[97: "a", 99: "c", 103: "g", 98: "b", 102: "f", 100: "d", 101: "e"]\n
nil
"[99: "c", 103: "g", 98: "b", 102: "f", 100: "d", 101: "e"]\n



"Value b is removed.\n"
```

图 5.24　字典中删除元素

5.3.4　遍历字典

遍历字典的元素可以在 for-in 循环中逐一访问键值对，并用元组保存。如图 5.25 所示，用元组变量(ascIICode, char)存储每一次循环取出的键值对，然后打印出来。

```
for (ascIICode,char) in ascIIDictChar {
    print("ascII code \(ascIICode) express char
        \(char) ")
}
```
(5 times)

图 5.25　遍历字典

5.3.5　字典的 keys 属性和 values 属性

字典类型还提供了 keys 和 values 属性，分别表示所有键和所有值的集合。如图 5.26 所示，分别遍历了字典的所有键和值的集合。

```
for ascIICode in ascIIDictChar.keys {
    print("keys:\(ascIICode);")
}
for char in ascIIDictChar.values {
    print("chars:\(char);")
}
```
(5 times)

(5 times)

图 5.26　字典的 keys 和 values 属性

为了便于使用 keys 属性和 values 属性，一般将其直接类型转换为一个数组变量，如图 5.27 所示。

```
let ascIICodeArray = Array(ascIIDictChar.keys)
let charArray = Array(ascIIDictChar.values)
```
[99, 103, 102, 100, 101]
["c", "g", "f", "d", "e"]

图 5.27　通过数组初始化集合

练习题

1. 定义一个字典型变量，表示大学里各学院的编号和学院名称。其中，编号为整型，学院名称为字符串型。对该字典型变量进行初始化，要求包含 6 个学院，编号从 1 开始。提示：这里编号为 key，学院名称为 value。

2. 用两种方法分别将 key 为 3 的元素 value 修改为 revisedSchool，key 为 4 的元素 value 修改为 updatedSchool，打印输出字典变量。

3. 书中没有介绍如何向字典中增加元素，请根据本章中介绍的相关知识尝试向字典中增加两个学院：编号 7 为 Mechanism、编号 8 为 Management。能否通过第 2 题中的方法实现，为什么？

4. 将第 3 题字典变量中的所有元素按照其编号和学院名称的对应关系逐个打印出来。

5. 将第 3 题字典变量中的 key 和 value 分别保存到数组 codeForSchool 和 schoolName 中，并通过 for-in 循环分别打印两个数组中的元素。

本章知识点

1. 3 种数组初始化的方法：从空数组开始逐步增加元素、直接赋值为一个具体的数组、将数组的所有元素都赋值为同一个初始值；数组的相加和累加运算；数组的下标运算；数组的插入与删除；数组的遍历；数组片段；数组元素的排序；数组元素交换位置以及检索特定元素。

2. 集合类型的初始化：从空集合开始、直接赋值；集合的为空判断和插入方法；

集合的删除。集合间的操作：交集、非交集、并集、相减。

3. 字典的两种声明方法和初始化方法；两种更新字典的方法：下标语法和 updateValue()方法；两种删除字典元素的方法：下标语法和 removeValue()；遍历字典；字典的 keys 属性和 values 属性。

参考代码

【5.1 节　数组】

第 1 题的参考代码-数组如图 5.28 所示。

```
var majorCompulsoryCourse = [String]()
majorCompulsoryCourse.append("Data Structure")
majorCompulsoryCourse.append("Computer Organizagtion")
majorCompulsoryCourse.append("Computer Networks")

var specializedOptionalCourse = ["iOS app development", "Swift
    Programming", "Artifical Intelligence"]

var generalKnowledgeCourse = [String](repeating: "Music", count: 3)
generalKnowledgeCourse[1] = "Painting"
generalKnowledgeCourse[2] = "Literature"
```

图 5.28　第 1 题的参考代码-数组

第 2 题的参考代码-数组如图 5.29 所示。

```
var allBookedCourse = majorCompulsoryCourse +
    specializedOptionalCourse + generalKnowledgeCourse
print(allBookedCourse)
```

图 5.29　第 2 题的参考代码-数组

第 3 题的参考代码-数组如图 5.30 所示。

```
majorCompulsoryCourse[0] = "Discrete Mathematics"
specializedOptionalCourse.insert("Experiment of Mobile Application
    Development", at: 3)
generalKnowledgeCourse.remove(at: 1)
```

图 5.30　第 3 题的参考代码-数组

第 4 题的参考代码-数组如图 5.31 所示。

```swift
allBookedCourse = majorCompulsoryCourse + specializedOptionalCourse
    + generalKnowledgeCourse
for (index, course) in allBookedCourse.enumerated() {
    print("No.\(index) course is \(course)")
}
```

图 5.31　第 4 题的参考代码-数组

第 5 题的参考代码-数组如图 5.32 所示。

```swift
let sliceOfAllBookedCourse = allBookedCourse[3...6]
var newArrayFromSlice = Array(sliceOfAllBookedCourse)
print(newArrayFromSlice)
```

图 5.32　第 5 题的参考代码-数组

第 6～8 题的参考代码-数组如图 5.33 所示。

```swift
if newArrayFromSlice.contains("Swift Programming") {
    print("It contains course: Swift Prgramming")
} else {
    print("No course: Swift Programming")
}
if newArrayFromSlice.contains("Swift") {
    print("It contains course: Swift")
} else {
    print("No course: Swift")
}
```

图 5.33　第 6～8 题的参考代码-数组

【5.2 节　集合】

第 1 题的参考代码-集合如图 5.34 所示。

```swift
var gradeOfTheory = Set<String>()
if gradeOfTheory.isEmpty {
    gradeOfTheory.insert("Fail")
    gradeOfTheory.insert("Pass")
    gradeOfTheory.insert("Common")
    gradeOfTheory.insert("Good")
    gradeOfTheory.insert("Excellent")
} else {
    print("Grade Set of Theory is not empty!")
}
```

图 5.34　第 1 题的参考代码-集合

第 2 题的参考代码-集合如图 5.35 所示。

```
var gradeOfExperiment: Set = gradeOfTheory
gradeOfExperiment.remove("Common")
gradeOfExperiment.remove("Good")
gradeOfExperiment.remove("Excellent")
if gradeOfExperiment.contains("Fail") && gradeOfExperiment.contains("Pass") {
    print("Grade Set of Experiment has two levels: Fail and Pass.")
} else {
    print("Grade Set of Experiment is lack of Fail or Pass level!")
}
```

图 5.35　第 2 题的参考代码-集合

第 3 题的参考代码-集合如图 5.36 所示。

```
print("All grades in Theory are: ")
for grade in gradeOfTheory {
    print("\(grade) ")
}

print("All grades in Experiment are: ")
for grade in gradeOfExperiment {
    print("\(grade) ")
}
```

图 5.36　第 3 题的参考代码-集合

第 4 题的参考代码-集合如图 5.37 所示。

```
gradeOfTheory.sorted()
print("All SORTED grades in Theory are: ")
for grade in gradeOfTheory {
    print("\(grade) ")
}

gradeOfExperiment.sorted()
print("All SORTED grades in Experiment are: ")
for grade in gradeOfExperiment {
    print("\(grade) ")
}
```

图 5.37　第 4 题的参考代码-集合

第 5 题的参考代码-集合如图 5.38 所示。

方法 subtracting()和 subtract()都是集合间的相减运算,不同之处在于,subtracting()不改变相减的两个集合的值,返回值为相减后的集合,而 subtract()用相减后的集合赋值给被减的集合,返回值为 void。

```
print("Intersection results are: ")
for grade in gradeOfTheory.intersection(gradeOfExperiment) {
    print("\(grade)")
}

print("Union results are: ")
for grade in gradeOfTheory.union(gradeOfExperiment) {
    print("\(grade)")
}

print("Subtract results are: ")
for grade in gradeOfTheory.subtracting(gradeOfExperiment) {
    print("\(grade)")
}
```

图 5.38　第 5 题的参考代码-集合

【5.3 节　字典】

第 1 题的参考代码-字典如图 5.39 所示。

```
var schoolOfUniversity = [Int:String]()
schoolOfUniversity = [1:"Material", 2:"Electronics", 3: "Astronics",
    4:"Dynamics", 5:"Aircraft", 6:"Computer"]
```

图 5.39　第 1 题的参考代码-字典

第 2 题的参考代码-字典如图 5.40 所示。

```
schoolOfUniversity[3] = "revisedSchool"
schoolOfUniversity.updateValue("updatedSchool", forKey: 4)
print(schoolOfUniversity)
```

图 5.40　第 2 题的参考代码-字典

第 3 题的参考代码-字典如图 5.41 所示。

```
schoolOfUniversity[7] = "Mechanism"
schoolOfUniversity[8] = "Management"
print(schoolOfUniversity)
```

图 5.41　第 3 题的参考代码-字典

第 4 题的参考代码-字典如图 5.42 所示。

```
print("Elements in Dictionary are as below:")
for (code, schoolName) in schoolOfUniversity {
    print("Code:\(code)---School Name:\(schoolName)")
}
```

图 5.42　第 4 题的参考代码-字典

第 5 题的参考代码-字典如图 5.43 所示。

```
var codeForSchool = Array(schoolOfUniversity.keys)
var schoolName = Array(schoolOfUniversity.values)
print("All codes are: ")
for code in codeForSchool {
    print("\(code )")
}
print("All school name are:")
for name in schoolName {
    print("\(name) ")
}
```

图 5.43　第 5 题的参考代码-字典

第 6 章 控 制 流

所谓控制流就是根据不同的控制条件决定程序的执行路径。本章主要介绍 Swift 中的各种循环控制语句,包括 for-in 循环、while 及 repeat-while 循环、if 条件语句、switch 条件语句以及控制流中的跳转语句。

6.1 for 循环

前面的章节已经用到 for-in 循环了,这里进行系统的讨论。

for 循环是指按照指定次数,重复执行一系列语句的操作。for 循环有两种形式,即 for-in 循环和 for 条件递增循环。

for-in 循环适用于遍历一个特定范围内的所有元素,如集合、数字范围、字符串或者数组。

图 6.1 遍历了 1~6 闭区间里的所有整数,这里的整型变量 i 在循环的声明语句中被隐式声明。每次循环时,i 被赋值 1~6 之间的数,在循环体内可以被引用。

图 6.1 for-in 遍历闭区间

在循环体中,如果不需要使用遍历的值,则使用下画线"_"代替具体的变量名。如图 6.2 所示,当用下画线代替变量名后,for-in 循环相当于重复执行一定次数循环体内的语句,而不需要遍历循环范围中的值参与循环体的执行。

```
for _ in 1...6 {
    print("Execute!")
}
```
(6 times)

图 6.2　下画线代替变量名

for-in 循环遍历数组、字典、集合的实例可参考前面相关的章节,这里不做重复介绍。

6.2　while 循环

while 循环就是重复执行一系列语句,直到条件语句的值为 false。

Swift 中有两种类型的 while 循环:第一种是 while 循环,在执行一系列语句前,先进行条件语句的判断。如果结果为 false,则结束循环;如果结果为 true,则继续执行循环体内的语句。第二种是 repeat-while 循环,先执行循环体内的一系列语句,然后进行条件语句的判断。如果结果为 false,则结束循环;如果结果为 true,则继续执行循环体内的语句。

while 循环的格式为

```
while condition {
    statements
}
```

repeat-while 循环的格式为

```
repeat {
    statements
} while condition
```

如图 6.3 所示,while 循环和 repeat-while 循环的判断条件相同,都是 i<5,执行的结果也一致。

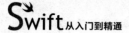

```
var i = 0
while i < 5 {                          0
    print("i = \(i)")                  (5 times)
    i += 1                             (5 times)
}
i = 0                                  0
repeat {
    print("i = \(i)")                  (5 times)
    i += 1                             (5 times)
} while i < 5
```

图 6.3 while 循环和 repeat 循环

6.3 if 条件语句

if 条件语句在前面章节中已经多次使用过,相信读者已经熟悉了。if 条件语句的格式为:

```
if  condition  {
statements
} else {
statements
}
```

其中,else 语句还可以继续嵌套新的 if-else 语句,嵌套的数量没有限制。例如:

```
if  condition1  {
statements
} else if condition2{
statements
} else {
statements
}
```

图 6.4 定义了一个字符串变量 season,用来保存当前的季节信息,并赋初值为"autumn"。通过层层嵌套 if-else 语句,分别判断 season 为四季中的哪一个,并打印相应的季节提示信息。

```
var season = "autumn"

if season == "spring" {
    print("All trees turn green.")
} else if season == "summer" {
    print("It's too hot.")
} else if season == "autumn" {
    print("Leaves are falling.")
} else {
    print("Snow will come!")
}
```

图 6.4 if 条件语句

6.4 switch 条件语句

switch 条件语句将一个值与若干个可能匹配的条件（或模式）进行比较，并执行第一个匹配成功的条件所对应的语句。如图 6.4 所示，当用 if-else 嵌套语句判断多种条件时，语句的复杂性提高，可读性下降。这种情况就可以用 switch 语句替代，在形式上会大大简化，且可读性较高。

switch 语句的格式如下：

```
switch someValue {
case value1: statementsFor1
case value2, 3: statementsFor23
default: statementsForDefault
}
```

6.4.1 匹配具体值

switch 语句中包含了多个 case 语句，每个 case 语句对应一个匹配的条件及相应的执行语句。注意，switch 语句中列出的各种匹配条件必须是完备的，即各种 case 情况必须包含 someValue 的所有可能取值。如果只想对部分特定情况进行比较和处理，可以用 default 语句处理其他没有出现在 case 中的可能情况（或值）。需要特别注意，在 switch 语句执行过程中，当第一次与 case 条件匹配后，执行该 case 中对应的语句，然后直接跳出 switch 语句块，继续执行后续语句，而其他 case 条件将被忽略。在

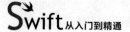

C 语言中将依次判断每个 case 条件，要忽略其他 case 条件，需要显式地使用 break 语句跳出 switch 块。

图 6.5 定义了一个枚举类型 month，包括 12 个月份的英文单词。接着，定义了一个 month 类型的变量 curMonth，表示当前的月份，并赋值为 month.February。通过 switch 语句对 curMonth 进行条件匹配，分别列出了 4 个匹配值。第一个条件为 month.January。第二个条件为 month.February。第三个条件为 month.April 或 month.May 或 month.June，只要和其中一个值相等，即匹配成功。第四个条件为默认值，如果和前面的条件都不匹配，则自动执行默认条件相应的语句。

```
enum month {
    case January
    case February
    case March
    case April
    case May
    case June
    case July
    case August
    case September
    case October
    case November
    case December
}
var curMonth = month.February

switch curMonth {
case month.January : print("the first
    month of a year")
case month.February : print("There is the
    most important festival in this month")
case month.April,month.May,month.June :
    print("month of the second season")
default : print("\(curMonth) doesn't match
    any case.")

}
```

图 6.5　switch 条件语句

6.4.2　匹配范围

switch 语句中 case 部分的条件也可以是一个范围。如图 6.6 所示，将上面的例子修改一下，用整型表示月份。case 语句分别对 1—3 月，4—6 月进行比较。

```
var curMonth = 5

switch curMonth {
case 1...3 : print("the first season")
case 4...6 : print("the second season")
default : print("the latter half of the year.")
}
```

图 6.6　case 语句

6.4.3　匹配元组值

switch 语句也可以对元组值进行匹配。如图 6.7 所示，classInfo 是一个课程信息的元组，包括开课年份和课程名称。通过 switch 语句对元组 classInfo 的信息进行比较，并根据不同情况打印出相应的课程信息。

```
var classInfo = (2015, "Computer Science")

switch classInfo {
case (2016, "Computer Science"): print(classInfo)
case (2016, _): print("2016's other class")
case (_,"Computer Science"): print("Computer Science
   Class in other year")
default : print("we don't care this class
   information")
}
```

图 6.7　switch 中的元组

6.4.4　值绑定

switch 语句中的 case 块可以进行值绑定，即将某个 case 语句匹配成功的值绑定到一个临时变量或常量上，在该 case 块的执行语句中可以使用该变量或常量。如图 6.8 所示，在 case 语句中分别将开课时间和课程名称的值绑定到常量 year 和 course 上，并在 case 块的执行语句中引用这两个常量。

```
classInfo = (2012, "Medicine")
switch classInfo {
case (let year, "Computer Science"): print("Year
   \(year)'s Computer Science Class")
case (2016, let course): print("2016's class
   \(course)")
case let (year,course): print("Class \(course) in
   year \(year)")
}
```

图 6.8　case 块的值绑定

6.4.5 where 语句

在值绑定的基础上，case 块的匹配语句可以引入 where 语句构造复杂的比较语句。如图 6.9 所示，通过 where 语句对特定条件进行筛选，匹配后执行相应语句。

```
switch classInfo {
case let (year,_) where year == 2015 :
    print("Class \(year) in year 2015")
case let (_,course) where course == "Medicine" :
    print("Class Medicine in year \(course)")
case let (year,course) :
    print("Class \(course) in year\(year)")
}
```

"Class Medicine in year Medicine\n"

图 6.9 where 条件筛选

6.5 控制转移语句

控制转移语句就是改变原有代码的执行顺序，实现代码跳转。这里介绍两个常用的控制转移语句：continue 语句和 break 语句。

continue 语句在循环语句中使用。当执行 continue 语句时，本次循环结束，直接跳转到下一次循环。如图 6.10 所示，在 for-in 循环语句中，当 i 等于 4 的时候执行 continue，那么本次循环就结束了，后续的打印语句将不会执行，程序直接跳转到下一次循环（i 等于 5）。

```
for i in 1...6 {
    if i == 4 {
        continue
    }
    print("No.\(i)")
}
```

(5 times)

图 6.10 continue 语句

break 语句可以用于循环语句中，也可以用于其他的控制流语句中。当执行 break 语句时，直接终止当前控制流，并跳转到控制流以外的后续语句处继续执行。

break 语句和 continue 语句在循环语句中应用时，差别是：break 语句执行后，终止全部后续的循环语句;continue 语句执行时,只结束当次循环语句的执行,并不跳出循环块。如图 6.11 所示,当 i 等于 4 的时候,终止后续所有的循环执行语句,直接跳出循环,继续执行循环块以外的代码。

```
for i in 1...6 {
    if i == 4 {
        break
    }
    print("No.\(i)")
}
```

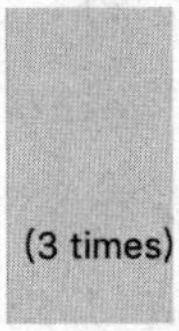

(3 times)

图 6.11　break 语句

练习题

1. 用 for 循环实现 1~100 的自然数的累加,并打印运算结果。

2. 在第 1 题的基础上,分别用 while 循环和 repeat-while 循环实现 1~100 的累加。

3. 用 if 条件语句实现:定义一个整型变量 score,用来表示考试成绩。当成绩小于 60 分时,打印出"Fail"的信息;当成绩大于或等于 60 分,但小于 70 分时,打印出"Pass"的信息;当成绩大于或等于 70 分,但小于 80 分时,打印出成绩为"Common"的信息;当成绩大于或等于 80 分,但小于 90 分时,打印出成绩为"Good"的信息;当成绩大于或等于 90 分时,打印出成绩为"Excellent"的信息。

4. 定义字符串变量 grade,用来记录成绩等级,即将第 3 题中打印输出的内容赋值给 grade。要求用 switch 条件语句实现。

5. 定义字符串变量 subject,用来记录科目名称,并赋值为"Math"。定义字符串变量 grade,赋值为"Excellent"。定义元组变量 subjectInfo,并赋值为(grade, subject)。用 switch 条件语句对如下几种情况进行匹配,并打印相应信息:grade 等于"Fail"、grade 等于"Pass"、grade 等于"Excellent"且 subject 等于"Math"、grade 等于"Excellent"且 subject 等于"Physics"。

6. 将元组 subjectInfo 赋值为("Pass","Math")。在第 5 题的基础上改写 switch 条件语句。当判断条件为:grade 等于"Fail"和 grade 等于"Pass"时,将科目名称绑定到常量 subject 上,并在打印语句中使用该常量。

7. 将元组 subjectInfo 赋值为("Excellent","Physics")。在第 6 题的基础上改写 switch 条件语句。当判断条件为：grade 等于"Excellent"且 subject 等于"Math"、grade 等于"Excellent"且 subject 等于"Physics"时，用 where 语句将这两种情况合并成一个条件，要求语义不变。

8. 定义变量 product，表示乘积，赋初值为 1。定义变量 count，赋初值为 1。用 while 循环实现当 count<100 时：count 自增 1，如果 product>100，则打印当前 count 和 product 的值，并用 break 跳出循环；否则，product 赋值为 product * count。循环结束后，打印当前 count 和 product 的值。

9. 将第 8 题中的 break 语句修改为 continue 语句，比较两段代码的差别，并说明原因。

本章知识点

1. for 循环的定义和用法。
2. while 循环和 repeat-while 循环的定义和用法。
3. if 条件语句的定义和用法。
4. switch 条件语句的定义以及条件语句的各种用法，包括匹配具体值、匹配范围、匹配元组值、值绑定以及 where 语句。
5. 两种控制转移语句的 break 和 continue 的用法与区别。

参考代码

第 1 题的参考代码如图 6.12 所示。

```
var sum = 0
for number in 1...100 {
    sum += number
}
print("Sum of 1+2+...+98+100 = \(sum)")
```

图 6.12　第 1 题的参考代码

第 2 题的参考代码如图 6.13 所示。

```
var number = 0
sum = 0
while number <= 100 {
    sum += number
    number += 1
}
print("Sum of 1+2+...+98+100 = \(sum)")

number = 0
sum = 0
repeat {
    sum += number
    number += 1
} while number <= 100
print("Sum of 1+2+...+98+100 = \(sum)")
```

图 6.13 第 2 题的参考代码

第 3 题的参考代码如图 6.14 所示。

```
var score = 88
if score < 60 {
    print("Fail")
} else if score < 70 {
    print("Pass")
} else if score < 80 {
    print("Common")
} else if score < 90 {
    print("Good")
} else {
    print("Excellent")
}
```

图 6.14 第 3 题的参考代码

第 4 题的参考代码如图 6.15 所示。

```
var grade = ""
switch score {
    case 0...59   : grade = "Fail"
    case 60...69  : grade = "Pass"
    case 70...79  : grade = "Common"
    case 80...89  : grade = "Good"
    case 90...100 : grade = "Excellent"
    default       : print("Unreasonable Score!")
}
```

图 6.15 第 4 题的参考代码

第 5 题的参考代码如图 6.16 所示。

```swift
var subject = "Math"
var grade = "Excellent"
var subjectInfo = (grade, subject)

switch subjectInfo {
case ("Fail", _): print("Fail")
case ("Pass", _): print("Pass")
case ("Excellent", "Math"): print("Math is Excellent")
case ("Excellent", "Physics"): print("Physics is Excellent")
default : print("Common or Good")
}
```

图 6.16　第 5 题的参考代码

第 6 题的参考代码如图 6.17 所示。

```swift
subjectInfo = ("Pass", "Math")
switch subjectInfo {
case ("Fail", let subject): print("Subject \(subject) is Fail")
case ("Pass", let subject): print("Subject \(subject) is Pass")
case ("Excellent", "Math"): print("Math is Excellent")
case ("Excellent", "Physics"): print("Physics is Excellent")
default : print("Common or Good")
}
```

图 6.17　第 6 题的参考代码

第 7 题的参考代码如图 6.18 所示。

```swift
subjectInfo = ("Excellent", "Physics")
switch subjectInfo {
case ("Fail", let subject): print("Subject \(subject) is Fail")
case ("Pass", let subject): print("Subject \(subject) is Pass")
case let (grade, subject) where (subject == "Math" || subject == "Physics")
    && grade == "Excellent": print("\(subject) is \(grade)")
default : print("Common or Good")
}
```

图 6.18　第 7 题的参考代码

第 8 题的参考代码如图 6.19 所示。

```
var product = 1
var count = 1
while count < 100 {
    count += 1
    if product > 100 {
        print("count is \(count), product is \(product).")
        break
    }
    product *= count
}
print("Cycle is stopped!")
print("count is \(count), product is \(product).")
```

图 6.19　第 8 题的参考代码

第 9 题的参考代码如图 6.20 所示。

```
product = 1
count = 1
while count < 100 {
    count += 1
    if product > 100 {
        print("count is \(count), product is \(product).")
        continue
    }
    product *= count
}
print("Cycle is stopped!")
print("count is \(count), product is \(product).")
```

图 6.20　第 9 题的参考代码

第 7 章 函 数

所谓函数,就是执行特定任务的独立的代码块。每个函数都有类型,函数的类型由参数类型和返回类型构成。本章介绍函数的定义和调用方法、函数形参、函数类型以及嵌套函数等。

7.1 函数的定义和调用方法

定义一个函数,首先要定义函数名。函数名是用来标识函数的,可以理解为一个函数的代号。从语法上看,函数名的定义并没有特殊要求,但为了提高代码的可读性,函数名要符合一定的规范。例如,一般函数名的第一个单词要小写,从第二个单词开始,每个单词的首字母要大写,具体可参考驼峰拼写法。另外,函数名要能够比较清楚简略地描述该函数所完成的任务,这可以帮助提高函数的可读性。

除了函数名,还要定义函数的传入值类型,即形参的类型,同时还要定义函数执行完毕后返回值的类型。

然后,还需要定义函数体,即一系列执行语句,用来完成特定的任务。

函数定义的格式为

```
func funcName(parameter: datatype) -> returnDatatype {
statements
return result
}
```

其中，func 为关键字；funcName 为函数名；parameter 为形参，可以有零到多个形参；datatype 为形参的数据类型；returnDatatype 为返回值的数据类型。花括号中的部分为函数体，由一系列执行语句 statements 组成，最后一条语句为返回结果。

如图 7.1 所示，函数名为 mulAdd，在函数名前面一定要有关键字 func。该函数有 3 个形参，分别代表两个乘数和一个加数，都是整型。函数的返回值也是整型，在返回值前面需要加上 -> 符号，表示该函数有返回值。花括号内是函数体，这里定义了乘加的具体运算过程，并将结果保存在常量 result 里。最后一条语句为返回运算结果，这条 return 语句和函数定义中的 -> 对应。函数定义中指明有返回值，就要有 return 语句。如果函数定义中没有返回值，则不需要 return 语句。

```
func mulAdd(mul1:Int,mul2:Int,add:Int) -> Int {
    let result = mul1*mul2 + add
    return result
}

var result = 0
result = mulAdd(mul1: 3, mul2: 5, add: 3)
print("The result of mulAdd is \(result)")
```

图 7.1 函数的定义和调用

使用一个函数完成一个特定的任务，称为函数调用。调用函数可以通过函数名，并根据形参的类型传入相应的值。当函数有返回值的时候，还要注意接收函数返回值的变量类型要和返回值类型一致。如图 7.1 所示，通过 mulAdd(mul1:3,mul2:5,add:3)调用函数，并将返回值保存在整型变量 result 中。

练习题

1. 编写一个打印首都信息的函数，函数名为 printCapitalInfo，有 3 个参数，分别为首都名 name、国家名 country 以及人口数 population。函数根据传入的参数打印出首都信息，如 Beijing is the capital of China and its population is 23 million。

2. 修改第 1 题的函数，将函数名修改为 calculateCapitalInfo，参数列表和 printCapitalInfo 相同，增加返回类型为元组型（String，Int），分别表示首都信息的字符串及其长度。

7.2 函数形参

7.2.1 无形参函数

函数形参的定义非常灵活,可以有 0 到多个形参。多个形参之间使用逗号分隔。在图 7.1 中,函数 mulAdd(mu1:Int,mul2:Int,add:Int)就有 3 个形参。图 7.2 改写了图 7.1 中的函数,将函数定义中的 3 个形参改为没有形参。不带形参的函数在调用的时候,不需要传入参数值,但是仍然需要在函数名后面有一对空的括号。

```
func mulAdd() -> Int {
    let mul1,mul2,add : Int
    mul1 = 3
    mul2 = 5
    add = 3
    return mul1*mul2 + add
}
result = mulAdd()
print("The result of mulAdd is \(result)")
```

图 7.2 函数形参

7.2.2 无返回值函数

函数不仅可以没有参数,也可以没有返回类型。当函数没有返回类型的时候,在 C 语言中称为过程。在 Swift 中,没有将无返回类型的函数作为一种特殊的情况考虑。如图 7.3 所示,继续改写前面的例子,将其变为一个无返回类型的函数。由于函数没有返回值,所以在函数定义的时候就没有 -> 符号了,在函数体中也不再需要 return 语句了。

```
func mulAdd(mul1:Int,mul2:Int,add:Int){
    print("The result of mulAdd is \(mul1*mul2 + add)")
}

mulAdd(mul1:3, mul2:5, add:3)
```

图 7.3 无返回值的函数

7.2.3 多个返回值函数

如果函数的返回值为多个,那么就需要用元组作为返回值类型。图7.4定义了一个气候的函数,根据传入参数城市返回这个城市的平均温度、天气以及风力信息。这里通过一个元组返回这3个信息。在函数体中,通过switch语句根据城市名称获取城市的气候信息。最后,通过return语句返回包含这3个气候信息的元组。

```swift
func climate(city:String)->(averageTemperature:Int,weather:String,wind:String) {
    var averageTemperature : Int
    var weather,wind : String
    switch city {
    case "beijing": averageTemperature = 25;weather = "dry";wind = "strong"
    case "shanghai": averageTemperature = 15;weather = "wet";wind = "weak"
    default : averageTemperature = 10; weather = "sunny"; wind = "normal"
    }
    return (averageTemperature,weather,wind)
}

var climateTemp = (0, "", "")
climateTemp = climate(city: "beijing")
```

图 7.4 多返回值函数

7.2.4 可变形参函数

除了确定个数的形参外,Swift还支持形参个数不确定的函数,即在函数定义的时候只知道形参的类型,并不知道形参的个数。只有在调用函数的时候,才能确定形参的个数,这种函数称为可变形参函数。

可变形参在定义的时候只需要定义参数名和参数类型,不同之处在于,参数类型后面要加上"…"符号,表示该形参为可变形参。可变形参在函数体内是以数组的形式存在的。图7.5定义了一个求和函数,形参为整型,个数要到调用的时候才知道。在函数体内,将形参numbers作为一个数组使用,通过for-in函数将其值循环累加。调用函数的时候,才能确定参数的值和个数。本例中,两次调用函数时,参数的个数分别为9和3。

```
func sum(numbers : Int...) -> Int{
    var result = 0
    for number in numbers {
        result = result + number
    }
    return result
}
sum(numbers: 1,2,3,4,5,6,7,8,9)
sum(numbers: 10,11,12)
```

图 7.5　求和的函数

7.2.5　inout 类型参数

前面介绍的形参只能在函数体内使用和改变，不会影响函数体以外传入的变量值。换言之，调用函数的时候，将函数外部变量作为参数传入函数体内的是变量值，而不是变量本身。当变量的值在函数体内发生变化时，函数体外的变量值并不会受到影响。如果要通过函数对传入的变量产生影响时，就需要在函数定义时在形参的类型中加上关键字 inout，表明该参数值的变化会影响传值给它的外部变量。这种 inout 类型的参数实际上是通过变量地址实现的。调用函数时，传递给函数中 inout 参数的不再是变量的值，而是变量的地址。在函数体中，当 inout 参数的值发生变化时，必然会导致指向同一地址的外部参数值发生改变。注意，调用函数时，不能将一个常量或者字面量传递给一个 inout 类型的参数。

图 7.6 定义了一个交换两个数的函数。在函数定义中，参数定义使用了关键字 inout，表示传入的不是值，而是地址。在函数体内执行两个数的值的交换。在函数调用时，传入的不是变量名，而是该变量的地址，这里使用了取址运算符 &。函数 swap() 执行完之后，打印结果显示外部变量 a 和 b 交换了值。

```
func swap(a:inout Int, b:inout Int) {
    let temp = a
    a = b
    b = temp
}

var a = 5
var b = 6
swap(a: &a, b: &b)
print("a is \(a)")
print("b is \(b)")
```

图 7.6　交换两个数的函数

练习题

1. 编写一个计算连乘的函数,函数名为 computeMultiply,没有参数,返回类型为 Int。该函数计算 1~10 连乘的积,并将结果返回。

2. 将计算连乘的函数修改为没有返回值的函数,函数名为 printMultiply。另外,函数要将计算结果打印输出。

3. 将计算连乘的函数修改为可变参数函数,增加一个 Int 型可变参数 multipliers,返回类型为 Int。该函数的参数为若干个整型数,以数组的形式保存在 multipliers 中。函数执行时,将 multipliers 中的数取出连乘,并返回结果。

4. 定义一个 Int 型数组变量 numbers,对其初始化,使其含有若干个元素。初始化后打印输出该数组。定义一个函数 sortArray(),其含有一个类型为 inout [Int]的参数 intArray,该函数负责对数组 intArray 进行排序。调用函数 sortArray(),其参数为 numbers。调用完毕后,再次打印输出该数组,比较函数调用前后的差别,并说明原因。

7.3 函数类型

每个函数都有特定的函数类型。函数类型由形参类型和返回值类型共同组成。如图 7.7 所示,add 的函数类型为(Int,Int)->Int,表示该函数有两个 Int 型的形参,返回值为一个 Int 型。helloWorld 的函数类型比较特殊,因为该函数既没有形参,也没有返回值,它的函数类型为()->(),表示该函数没有形参,返回值为 void,相当于一个空元组。

```
func add(a: Int,b: Int) ->Int{
    return a+b
}

func helloWorld() {
    print("Hello world!")
}
```

图 7.7 函数类型

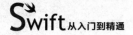

7.3.1 函数类型变量

函数类型可以像其他数据类型一样使用。图 7.8 定义了两个变量 mathOperation 和 sayOperation。第一个变量的类型是(Int,Int)->Int,这是一个函数类型,表示变量 mathOperation 为一个函数变量,可以将函数类型为(Int,Int)->Int 的函数赋值给它。第二个变量的类型是()->(),这也是一个函数类型,表示变量 sayOperation 为一个函数变量,可以将函数类型为()->()的函数赋值给它。在这两个变量的定义语句中,将其赋值为上例中定义的两个函数 add()和 helloWorld()。这两个已知函数的类型与变量 mathOperation、sayOperation 定义的函数类型匹配,因此赋值是合法的。

函数类型和其他类型一样,也可以用来给一个变量赋值。变量的类型不需要显式地指出,编译器会通过类型推断得到变量类型。在图 7.8 中,直接给变量 operation 赋值为函数名,可以看到左侧显示,系统根据类型推断得到函数类型为(Int,Int)->Int。

```
var mathOperation : (Int,Int)->Int = add         (Int, Int) -> Int
var sayOperation : ()->() = helloWorld           () -> ()

mathOperation(5,6)                               11
sayOperation()

var operation = add                              (Int, Int) -> Int
operation(6, 7)                                  13
```

图 7.8　函数类型变量

7.3.2 函数类型的参数

函数类型还可以作为另一个函数的参数类型使用。这一点大大增加了函数定义的灵活性,可以在调用函数的时候,再根据需要决定哪个函数被调用,而不需要在编写代码的时候就确定。图 7.9 定义了两个简单的数学运算函数 add()和 sub(),分别表示两个数的加法运算和减法运算。这里默认认为 a 是大于 b 的,否则进行减法运算时会出现负数,在某些情况下出现负数是不允许的。然后,定义了函数 printResult(),该函数的形参中包含一个函数类型的参数 operation,该参数的类型为函数类型(Int,Int)->Int。因此,调用函数 printResult()时,可以将一个类型匹配的函数传给这个参数。在 printResult 函数体中,根据 a、b 的大小决定调用 operation 时参数的顺序,从而保证不会出现小数减大数的情况。在函数体外,调用 printResult()函数,并将减法函数 sub() 传入给形参 operation。

```
func add(a: Int,b: Int) ->Int{
    return a + b
}
func sub(a: Int,b: Int) ->Int{
    return a - b
}
func printResult(operation: (Int,Int)->Int, a:Int, b:Int) {
    let result : Int
    if a>b {
        result = operation(a,b)
    } else {
        result = operation(b,a)
    }
    print("the result is \(result)")
}

printResult(operation: sub, a: 3, b: 9)
```

图 7.9 函数类型作为函数的参数

7.3.3 函数类型的返回值

前面介绍了将函数类型作为另一个函数的参数使用。同样，函数类型也可以作为另一个函数的返回值类型使用。图 7.10 定义了一个返回值类型为函数类型(Int, Int)->Int 的函数 mathOperation()，该函数根据传入的形参 op 的值确定返回什么函数。当 op 等于 sub 时，返回减法函数，否则返回加法函数。函数调用时，传入的形参值为 sub，所以返回 sub() 函数。因此，常量 result 的值为 sub() 函数。当调用 result 时，效果与直接调用 sub 相同。

```
func mathOperation(op: String) -> (Int,Int)->Int {
    if op == "sub" {
        return sub
    }else {
        return add
    }
}

let result = mathOperation(op: "sub")
result(6,3)
```

图 7.10 函数类型作为函数的返回类型

练习题

1. 定义两个函数类型的变量。其中,变量 printAction 的类型为()->()。将 7.2 节练习题中编写的函数 printMultiply()赋值给变量 printAction。变量 mathOperation 的类型为(Int…)->Int。将 7.2 节练习题中编写的函数 computeMultiply()赋值给变量 mathOperation。通过这两个函数类型的变量调用函数。

2. 定义两个函数,函数名分别为 multiply 和 divide,函数类型均为(Int,Int)->Int。其中,multiply 返回两个数相乘的结果,divide 返回两个数相除的结果。在此基础上定义名为 printMathOperation 的函数,函数类型为((Int,Int)->Int,Int,Int)->()。该函数含有 3 个参数:一个函数、两个整数。它调用传入的函数对两个整数进行算术运算。当传入函数为 divide 时,始终为大数除以小数。将 multiply 和 divide 作为参数分别调用函数 printMathOperation()。

3. 在第 2 题的基础上定义一个返回值类型为函数的函数。函数名为 multiplyOrDivideOperation,含有一个字符串型参数 op,返回值的类型为(Int,Int)->Int。当 op 等于 Multiply 时,返回函数 multiply(),否则返回函数 divide()。定义一个函数类型的变量 operation,调用函数 multiplyOrDivideOperation(),参数为 Multiply。再以 operation 和任意两个整数作为参数,调用函数 printMathOperation()。

7.4 嵌套函数

在函数体内还可以定义新的函数,称之为嵌套函数。嵌套函数只能在函数体内使用,对函数体外是不可见的。如图 7.11 所示,改写前面的例子,将加、减法运算函数的定义移到函数 printResult()内。在函数 printResult()中,可以直接引用函数体中的嵌套函数 add()和 sub()。

练习题

用嵌套函数改写 7.3 节中的函数 multiplyOrDivideOperation(),参数修改为 3 个,分别为字符串型 op、整型 num1 和 num2,返回值的类型为 Int。要求根据参数 op 的值调用子函数 multiply()或 divide()。

```
func printResult(a:Int, b:Int) {
    func add(a: Int,b: Int) ->Int{
        return a + b
    }
    
    func sub(a: Int,b: Int) ->Int{
        return a - b
    }
    
    let result : Int
    if a>b {
        result = sub(a: a, b: b)
    } else {
        result = add(a: b, b: a)
    }
    print("the result is \(result)")
}

printResult(a: 6, b: 3)
```

图 7.11 嵌套函数实例

本章知识点

1. 函数的定义和调用。

2. 函数参数的相关概念，包括无形参函数、无返回值函数、多个返回值函数、可变参数函数以及 inout 类型函数参数。

3. 函数类型的应用：函数类型的变量、函数类型的参数、函数类型的返回值。

4. 嵌套函数的使用。

参考代码

【7.1 节 函数的定义和调用】

第 1 题的参考代码-函数的定义和调用如图 7.12 所示。

```
func printCapitalInfo(name: String, country: String, population: Int) {
    print("\(name) is the capital of \(country) and its population is
        \(population) million.")
}

printCapitalInfo(name: "Beijing", country: "China", population: 23)
```

图 7.12 第 1 题的参考代码-函数的定义和调用

第 2 题的参考代码-函数的定义和调用如图 7.13 所示。

```swift
func calculateCapitalInfo(name: String, country: String, population: Int) ->
    (String, Int) {
    let capitalInfo = name + " is the capital of " + country + " and its
        population is " + String(population) + " million."
    let lengthOfInfo = capitalInfo.count
    return (capitalInfo, lengthOfInfo)
}

let capitalInfo = calculateCapitalInfo(name: "Beijing", country: "China",
    population: 23)
print(capitalInfo)
```

图 7.13　第 2 题的参考代码-函数的定义和调用

【7.2 节　函数形参】

第 1 题的参考代码-函数形参如图 7.14 所示。

```swift
func computeMultiply() -> Int {
    var result = 1
    for i in 1...10 {
        result *= i
    }
    return result
}

var result = 0
result = computeMultiply()
```

图 7.14　第 1 题的参考代码-函数形参

第 2 题的参考代码-函数形参如图 7.15 所示。

```swift
func printMultiply() {
    var result = 1
    for i in 1...10 {
        result *= i
    }
    print("10! = \(result)")
}

printMultiply()
```

图 7.15　第 2 题的参考代码-函数形参

第 3 题的参考代码-函数形参如图 7.16 所示。

```swift
func computeMultiply(multipliers : Int...) -> Int {
    var result = 1
    for i in multipliers {
        result *= i
    }
    return result
}

let product = computeMultiply(multipliers: 1,3,5,7,9)
```

图 7.16　第 3 题的参考代码-函数形参

第 4 题的参考代码-函数形参如图 7.17 所示。

```swift
var numbers = [15,3,90,2,0,8,10,12]
print("unsorted array is: \(numbers)")

func sortArray(intArray: inout [Int]) {
    intArray.sort()
}

sortArray(intArray: &numbers)
print("sorted array is: \(numbers)")
```

图 7.17　第 4 题的参考代码-函数形参

【7.3 节　函数类型】

第 1 题的参考代码-函数类型如图 7.18 所示。

```swift
var printAction : ()->() = printMultiply
var mathOperation: (Int...)->Int = computeMultiply

printAction()
let returnValue = mathOperation(2,4,8)
```

图 7.18　第 1 题的参考代码-函数类型

第 2 题的参考代码-函数类型如图 7.19 所示。

```swift
func multiply(a: Int, b: Int) -> Int {
    return a * b
}

func divide(a: Int, b: Int) -> Int {
    return a / b
}

func printMathOperation(operation: (Int, Int)->Int, a: Int, b: Int) {
    let result : Int
    if a > b {
        result = operation(a,b)
    } else {
        result = operation(b,a)
    }
    print("the result of operation between \(a) and \(b) is \(result)")
}

printMathOperation(operation: multiply, a: 28, b: 5)
printMathOperation(operation: divide, a: 28, b: 5)
```

图 7.19　第 2 题的参考代码-函数类型

第 3 题的参考代码-函数类型如图 7.20 所示。

```swift
func multiplyOrDivideOperation(op: String) -> (Int, Int)->Int {
    if op == "Multiply" {
        return multiply
    } else {
        return divide
    }
}

var operation = multiplyOrDivideOperation(op: "Multiply")
printMathOperation(operation: operation, a: 9, b: 108)

operation = multiplyOrDivideOperation(op: "Divide")
printMathOperation(operation: operation, a: 9, b: 108)
```

图 7.20　第 3 题的参考代码-函数类型

【7.4 节　嵌套函数】

参考代码-嵌套函数如图 7.21 所示。

```
func multiplyOrDivideOperation(op: String, num1: Int, num2: Int)
    -> Int {

    func multiply(a: Int, b: Int) -> Int {
        return a * b
    }

    func divide(a: Int, b: Int) -> Int {
        return a / b
    }


    if op == "Multiply" {
        return multiply(a:num1, b:num2)
    } else {
        if num1 > num2 {
            return divide(a:num1,b:num2)
        } else {
            return divide(a:num2,b:num1)
        }
    }
}

var result = multiplyOrDivideOperation(op: "Divide", num1:5,
    num2:15)
```

图 7.21 参考代码-嵌套函数

第 8 章 闭 包

闭包是一种功能性自包含模块,可以捕获和存储上下文中任意常量和变量的引用。

闭包有 3 种形式,分别如下。

(1) 全局函数:有名字但不会捕获任何值的闭包。

(2) 嵌套函数:有名字并可以捕获其封闭函数域内值的闭包。

(3) 闭包表达式:没有名字但可以捕获上下文中变量和常量值的闭包。

嵌套函数是一种在复杂函数中命名和定义自包含代码块的简洁方式,而闭包表达式则是一种利用内联闭包方式实现的更简洁的方式。全局函数和嵌套函数是特殊形式的闭包,在函数一章中已经介绍过。本章介绍闭包表达式,后文中提到的闭包都指闭包表达式。

8.1 闭包表达式

所谓闭包,就是将其变量和常量封装在自己的作用域内。闭包可以访问、存储和操作上下文中的任何变量和常量,这个过程称为变量或常量被闭包捕获。

8.1.1 闭包的定义

闭包表达式的格式为

{ (parameters) ->returnType in

```
statements
}
```

闭包表达式的参数可以为常量和变量,也可以使用 inout 类型,但不能提供默认值。参数列表中的最后一个参数可以是可变参数。闭包表达式也可以使用元组作为参数和返回值。

和函数一样,闭包也可以赋值给变量或常量。图 8.1 定义了一个闭包类型的变量 theClosure,并赋值为一个类型为(String, String)->Bool 的闭包。该闭包的功能是比较两个字符串参数的 ASCII 码值大小,并以布尔值返回结果。

```
var theClosure = { (s1: String, s2: String) ->
    Bool in return s1 > s2 }
```
(String, String) -> Bool

图 8.1　闭包的定义

闭包是一种特殊的函数,和函数一样都有参数列表和返回类型。不同之处在于,在函数中,这些元素都在花括号外,而在闭包中,这些元素都在花括号内,并且用关键字 in 连接。

调用闭包时,除了不需要提供参数名外,和函数没有差别,如图 8.2 所示。

```
let compareResult = theClosure("Zoo", "Park")    true
```

图 8.2　闭包的调用

8.1.2　闭包类型参数

Swift 中的数组类型有一个 sort()方法,其功能是对数组中的值进行排序。排序的规则由一个已知的闭包提供。该方法的返回值为一个排过序的新数组。sort()方法只有一个参数,为闭包类型。该闭包有两个参数,参数类型与数组元素的类型一致。闭包的返回值为布尔型。当返回值为 true 时,表示第一个元素和第二个元素位置保持不变。当返回值为 false 时,表示第一个元素应该和第二个元素交换位置。

例如,用方法 sort()对一个 String 类型的数组按照字母降序排列,数组的初始值为["Beijing","Shanghai","Guangzhou","Hangzhou","Suzhou"]。那么,sort()方法

的参数类型就为(String,String)->Bool 的闭包。

　　首先,用全局函数作为 sorted()方法的闭包类型参数。如图 8.3 所示,定义函数 exchange(),用来判断两个字符串的值,并将比较结果以布尔值形式返回。如果第一个字符串大于第二个字符串,返回 true,表示需要交换两个字符串在数组中的位置。反之,返回 false,表示不需要交换两个字符串在数组中的位置。将函数名 exchange 作为参数传给数组 cityArray 的方法 sorted()。方法 sorted()的返回值保存到变量 descendingArray 中。

```swift
func exchange(s1:String, s2:String)->Bool {
    return s1 > s2
}

let cityArray = ["Beijing","Shanghai","Guangzhou","Hangzhou","Suzhou"]

var descendingArray = cityArray.sorted(by: exchange)
print(descendingArray)
```

图 8.3　exchange()函数

　　然后,再用闭包作为 sorted()方法的闭包类型参数。如图 8.4 所示,将 exchange()函数改写为闭包,并将其作为参数传给数组 cityArray 的方法 sorted()。用闭包和用函数的效果完全一致,但是简洁很多。

```swift
var descengdingArrayByClosures = cityArray.sorted(by: {(s1:String, s2:String)->Bool
    in return s1 > s2})
print(descengdingArrayByClosures)
```

图 8.4　闭包表达式

　　在此基础上,还可以进一步简化闭包,如图 8.5 所示。由于 sorted()方法的参数类型是明确的,即闭包类型(String,String)->Bool。如果没有闭包内相关类型的声明,编译器也可以根据上下文推断闭包中的参数类型和返回值类型。因此,可以去掉闭包中的相关类型声明,即简化为闭包{s1,s2 in return s1>s2}。其次,还可以去掉 return 关键字,隐式表示返回结果,即简化为闭包{s1,s2 in s1>s2}。另外,Swift 还提供了一种参数简写的方式,即 $0 表示第一个参数,$1 表示第二个参数。采用这种参数简写的方式连参数声明都多余了,只要保留参数之间的运算关系即可,即简化为闭包{ $0> $1}。最后,由于运算表达式中参数的出现次序默认为依次出现,即先出现$0,后出现$1,则还可以简化掉参数本身,只保留运算符,即简化为闭包{>}。上述闭包的各种简化的执行效果都是一致的。需要注意的是,代码的简化会使可读性下降。因此,在什么情况下使用闭包,闭包简化到什么程度,是需要慎重考虑的。

```
var descendingArray2 = cityArray.sorted(by: {s1,s2 in return s1>s2})
print(descendingArray2)

var descendingArray3 = cityArray.sorted(by: {s1,s2 in s1>s2})
print(descendingArray3)

var descendingArray4 = cityArray.sorted(by: {$0 > $1})
print(descendingArray4)

var descendingArray5 = cityArray.sorted(by: >)
print(descendingArray5)
```

图 8.5　闭包表达式的简化

8.1.3　无参数无返回值的闭包

当闭包的类型为()->Void 时,称之为无参数无返回值闭包。一对空括号表示没有参数,但不能省略。Void 表示无返回值,也不能省略,否则编译器无法断定其为闭包。定义一个变量为闭包类型时,闭包的具体类型可以显式声明,也可以通过赋初始值隐式声明,两者的效果是一样的。图 8.6 定义了两个无参数无返回值的闭包变量,其闭包类型分别采用显式声明和隐式声明方式。这里的闭包调用和函数调用完全一样。

```
var voidClosureWithTypeDeclaration : () -> Void =          () -> ()
    {
    print("A closure without parameters and                ()
        return value")
    }

var voidClosure = {                                        () -> ()
    print("Another clousre without parameters and          ()
        return value")
    }
voidClosureWithTypeDeclaration()
voidClosure()
```

图 8.6　无参数无返回值的闭包

8.1.4 尾随闭包

如果闭包表达式是函数的最后一个参数,则可以使用尾随闭包的方式增加代码的可读性。

尾随闭包是将整个闭包表达式从函数的参数括号里移到括号外面的一种书写方式。图 8.7 定义了函数 mathCompute(),其功能是根据不同的运算符号进行相应的运算。该函数有 4 个参数,其中最后一个参数为闭包类型,即(Int, Int)->Int。调用函数 mathCompute()时,给出了两种写法:第一种是将整个闭包作为参数的写法;第二种是尾随闭包的写法。两种写法的执行效果一样,但采用尾随闭包的写法会明显提升可读性,尤其是当闭包的代码较多的时候。

```
func mathCompute(opr:String,n1:Int,n2:Int,compute:(Int,Int)->Int) {
    switch opr {
    case "+": print("\(n1)+\(n2) = \(compute(n1,n2))")
    case "-": print("\(n1)-\(n2) = \(compute(n1,n2))")
    case "*": print("\(n1)*\(n2) = \(compute(n1,n2))")
    case "/": print("\(n1)/\(n2) = \(compute(n1,n2))")
    default : print("not support this operator")
    }
}

mathCompute(opr: "+", n1: 9, n2: 3, compute: {(n1:Int,n2:Int)->Int in n1 + n2})
mathCompute(opr: "*", n1: 9, n2: 3){(n1:Int,n2:Int)->Int in n1 * n2}
```

图 8.7 尾随闭包实例

练习题

1. 定义一个闭包变量 compareClosure,并将一个闭包赋值给它。该闭包对两个整型参数进行比较,并返回比较结果(布尔型)。调用 compareClosure 两次,参数分别为(8,10)和(10,8)。

2. 定义一个数组常量 numbers,并赋初始值为[88,128,9,3,66,16,8]。第一次调用 numbers 的数组类型方法 sorted(),要求将第 1 题编写的闭包表达式作为其参数。第二次调用该方法,要求运用本节介绍的方法,将该闭包表达式化为最简形式。第三次调用该方法,要求将闭包变量 compareClosure 作为参数。比较 3 次调用的结果是否有差别,并说明原因。

3. 定义一个无返回值无参数的闭包。要求计算 1…100 的累加值,并打印。

4. 将第 2 题中的第一次调用 numbers.sorted() 方法的语句修改为尾随闭包的写法。

8.2 闭包的应用

8.2.1 捕获闭包作用域的变量

闭包可以捕获其作用域中的任何变量。图 8.8 定义了一个变量 number，赋初始值为 100，然后又定义了一个闭包类型的常量 minusNumber，该闭包类型为无返回无参数，即 ()->Void。闭包体中只有一条语句，即变量 number 自减 1。变量 number 在闭包的作用域内，因此闭包可以捕获该变量。在闭包体中，捕获到变量 number 后对其值进行修改，修改后的结果会体现到闭包体外，通过 3 次调用闭包 minusNumber 后，number 变为 97 就可以说明这一点。

```
var number = 100

let minusNumber = {
    number -= 1
}

minusNumber()
minusNumber()
minusNumber()
print(number)
```

100

() -> ()
(3 times)

"97\n"

图 8.8 捕获闭包作用域的变量

利用闭包可以捕获其作用域中的变量，不同闭包实例的作用域相互独立。图 8.9 定义了一个函数 computeNumberClosure()，其返回值为闭包类型。在函数体内定义了一个变量 number 和一个闭包 minusNumber。该闭包对变量 number 进行自减运算，并返回该变量。函数的返回值为闭包 minusNumber。在函数体外，定义了两个闭包类型的常量 firstCountDown 和 secondCountDown，并分别赋值为 computeNumberClosure()，即这两个常量各自得到一个独立的闭包实例 minusNumber。因此，闭包 firstCountDown 和 secondCountDown 有各自的作用域且相互独立。调用 firstCountDown() 两次后，其作用域中的 number 值为 98；而调用 secondCountDown() 4 次后，其作用域中的 number 值为 96。由此可见，两个闭包的执行是相互独立的。

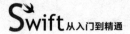

```
func computeNumberClosure() -> ()->Int {          
    var number = 100                              (2 times)
    let minusNumber: ()->Int = {                  (2 times)
        number -= 1                               (6 times)
        return number                             (6 times)
    }
    return minusNumber                            (2 times)
}

let firstCountDown = computeNumberClosure()       () -> Int
let secondContDown = computeNumberClosure()       () -> Int

firstCountDown()                                  99
secondContDown()                                  99
secondContDown()                                  98
secondContDown()                                  97
firstCountDown()                                  98
secondContDown()                                  96
```

图 8.9　闭包的作用域

利用闭包还可以方便地对集合类型变量进行各种遍历操作，包括修改元素值、过滤出符合特定条件的元素等。

8.2.2　方法 forEach()

集合类型的方法 forEach() 负责遍历集合中的每一个元素，而对元素的操作可以通过闭包给出。如图 8.10 所示，数组变量 numbers 含有 6 个元素，将闭包作为参数传给 numbers 数组的 forEach() 方法，从而实现依次打印数组中的每一个元素及其立方值。

```
22  var numbers = [1,2,3,4,5,6]                   [1, 2, 3, 4, 5, 6]
23  numbers.forEach{                              [1, 2, 3, 4, 5, 6]
24      print("The Cube of \($0) is \($0*$0*$0)") (6 times)
25  }
```

```
The Cube of 1 is 1
The Cube of 2 is 8
The Cube of 3 is 27
The Cube of 4 is 64
The Cube of 5 is 125
The Cube of 6 is 216
```

图 8.10　方法 forEach()

8.2.3 方法 filter()

方法 filter() 负责按照一定的条件将集合中的元素过滤出来,并形成一个新的集合,过滤的条件由闭包类型的参数提供。如图 8.11 所示,在上例的基础上定义一个常量 filteredNumbers。调用 numbers 数组的方法 filter(),将其返回值赋给 filteredNumbers。方法 filter() 的参数为闭包类型,该闭包定义过滤的条件为元素值大于 3。方法 filter() 的返回值为一个数组,该数组为 numbers 的子数组且每个元素值都大于 3。

```
let filteredNumbers = numbers.filter{
    return $0 > 3
}
```
[4, 5, 6]
(6 times)

图 8.11 方法 filter()

8.2.4 方法 map()

方法 map() 负责遍历集合中的每一个元素,对其进行操作,并形成一个新的集合,对元素的操作由闭包类型的参数提供。如图 8.12 所示,在上例的基础上定义一个常量 doubledNumbers。调用 numbers 数组的方法 map(),将每个元素乘以 2 后形成一个新的数组。

```
let doubledNumbers = numbers.map{
    return $0 * 2
}
```
[2, 4, 6, 8, 10, 12]
(6 times)

图 8.12 方法 map()

8.2.5 方法 reduce()

方法 reduce() 有两个参数:第一个参数为整型,作为起始值;第二个参数为闭包类型。这个闭包有两个参数:一个为当前值;另一个为集合中的一个元素值。闭包的返回值作为传给自己的当前值参数。如图 8.13 所示,在上例的基础上定义一个常量 sumOfNumbers。调用 numbers 数组的方法 reduce(),起始值参数为 0,将其与数组中的每个元素值相加,每次相加的结果作为下次相加时的当前值。相加的最终结果作为方法 reduce() 的返回值,赋值给 sumOfNumbers。

```
let sumOfNumbers = numbers.reduce(0){
    return $0 + $1
}
```

21
(6 times)

图 8.13　方法 reduce()

练习题

1. 定义一个函数,用于模拟马拉松运动员比赛的情况。函数名为 marathonRace,有两个参数,分别为:speed 表示运动员每分钟前进多少米,为整型;racer 表示运动员的姓名,为字符串型。函数的返回值为()->Int。在函数体中用变量 remainDistance 表示剩余的距离,马拉松全长为 421 950m。在函数体中定义一个闭包 racing,每分钟从 remainDistance 中减去 speed 米,并返回当前剩余的距离。另外,定义两个常量 runnerJack 和 runnerTom,用来模拟运动员 Jack 和 Tom,他们每分钟前进的速度分别为 280m 和 315m。调用函数 marathonRace(),将其返回值赋给 runnerJack 和 runnerTom。利用循环语句模拟计时器,跟踪两个运动员每分钟前进的情况,并打印第一个达到终点的运动员名字。

2. 定义一个字符串型数组变量 capitals,用来保存各国的首都名称。初始化该数组,向其中加入一些你所知道的首都英文名。编写尾随闭包,将其作为参数赋值给数组类型的相应方法,具体实现如下要求:①通过 forEach() 方法打印数组中的每个元素。②通过 filter() 方法将数组中含有字母"o"的元素过滤出来,并形成一个新的数组。③通过 map() 方法将数组中所有的元素都变成小写,并形成一个新的数组。④通过 reduce() 方法将数组中的所有元素连接成一个字符串,每个元素间要有分隔符。在字符串首要增加提示信息"Capitals are:"。

本章知识点

1. 3 种类型的闭包:全局函数;嵌套函数;闭包表达式及其简化;尾随闭包的定义和写法。

2. 闭包的应用:通过闭包定制集合类型数据的迭代操作(包括 forEach、filter、

map、reduce)。

参考代码

【8.1 节 闭包表达式】

第 1 题的参考代码-闭包表达式如图 8.14 所示。

```
var compareClosure = {
    (a: Int, b: Int) -> Bool in
    return a > b
}

compareClosure(8, 10)
compareClosure(10, 8)
```

```
(Int, Int) -> Bool

(13 times)



false
true
```

图 8.14 第 1 题的参考代码-闭包表达式

第 2 题的参考代码-闭包表达式如图 8.15 所示。

```
let numbers = [88,128,9,3,66,16,8]

var sortedNumbers = numbers.sorted(by:
    {(a: Int, b: Int) -> Bool in return a
    > b})
sortedNumbers = numbers.sorted(by: >)
sortedNumbers = numbers.sorted(by:
    compareClosure)
```

```
[88, 128, 9, 3, 66, 16, 8]

(12 times)


[128, 88, 66, 16, 9, 8, 3]

[128, 88, 66, 16, 9, 8, 3]
```

图 8.15 第 2 题的参考代码-闭包表达式

第 3 题的参考代码-闭包表达式如图 8.16 所示。

```
var voidClosure = {
    var sum = 0
    for i in 1...100 {
        sum += i
    }
    print("1 + ... + 100 = \(sum)")
}
voidClosure()
```

```
() -> ()
0


(100 times)

"1 + ... + 100 = 5050\n"
```

图 8.16 第 3 题的参考代码-闭包表达式

第 4 题的参考代码-闭包表达式如图 8.17 所示。

```
    sortedNumbers = numbers.sorted(){(a: Int,   (12 times)
        b: Int) -> Bool in return a > b}
```

图 8.17 第 4 题的参考代码-闭包表达式

【8.2 节 闭包的应用】

第 1 题的参考代码-闭包的应用如图 8.18 所示。

```
func marathonRace(speed: Int, racer: String) -> () -> Int {
    var remainDistance = 421950
    let racing: () -> Int = {
        remainDistance -= speed
        return remainDistance
    }
    return racing
}

let runnerJack = marathonRace(speed: 280, racer: "Jack")
let runnerTom  = marathonRace(speed: 315, racer: "Tom")


for _ in 1...1500 {
    if runnerJack() <= 0 {
        print("Jack arrived")
        break
    }
    if runnerTom() <= 0 {
        print("Tom arrived")
        break
    }
}
```

图 8.18 第 1 题的参考代码-闭包的应用

第 2 题的参考代码-闭包的应用如图 8.19 所示。

```
var capitals = ["Beijing","Tokyo","Moscow","Berlin","Washington","London"]

capitals.forEach{ print($0)}

let filteredCapitals = capitals.filter{
    return $0.contains("o") }

let lowerCasedCapitals = capitals.map{
    return $0.lowercased()
}

let capitalString = capitals.reduce("Capitals are: "){
    return $0 + "," + $1
}
```

图 8.19 第 2 题的参考代码-闭包的应用

第 2 篇

高 级 篇

　　高级篇由 7 个章节构成，分别为枚举型、结构体与类、属性、方法、类、协议以及泛型，都是面向对象的相关知识。Swift 中的面向对象在概念上与 C 语言没有差别，但在具体的实现细节上有不少独特之处。例如，结构体可以有方法，属性有存储属性、计算属性、属性观察器等。对于已经有面向对象概念的读者，要特别注意 Swift 中面向对象概念的特点。

第 9 章 枚举型

枚举型是一种普通的类型，该类型在一个特定的集合中取值。集合中的元素称为枚举成员。枚举型是类型安全的，当向一个枚举型变量或常量赋非法值时，编译器能在输入后立即报错。

Swift 中枚举型的功能非常强大，除了拥有 C 语言中枚举型的特点外，它还有很多面向对象的能力。枚举成员都可以有一个与其对应的值，即关联值。枚举成员的关联值可以为字符串、字符、整型或者浮点型等。枚举成员也可以有一个对应的原始值。另外，枚举型还具有一些类的特征，如计算属性、实例方法、构造器以及协议等。而在 C 语言中，枚举型则是一种基本数据类型，类似于整型、字符串型等。

9.1 枚举型的定义

定义枚举型要使用关键字 enum，具体格式如下：

```
enum Name {
case ValueName
}
```

枚举型是一个自定义的类型。枚举型的名字必须以大写字母开头。枚举成员前面必须使用关键字 case。图 9.1 定义了名为 MonthFull 的枚举型，每个枚举成员都用一个不同月份的英文表示。

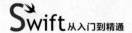

```
enum MonthFull {
    case January
    case February
    case March
    case April
    case May
    case June
    case July
    case August
    case September
    case October
    case November
    case December
}
```

图 9.1 枚举型实例

9.1.1 枚举型的紧凑写法

多个枚举成员也可以出现在同一行中,但要用逗号隔开,如图 9.2 所示。这种写法比较紧凑。

```
enum Month {
    case January,February,March,April,May,June,
        July,August,September,October,November,December
}
```

图 9.2 枚举型的紧凑写法

9.1.2 枚举型的使用

定义好枚举型 Month 后,就可以使用了。如图 9.3 所示,在声明一个 Month 类型的变量时,可以直接通过 Month.August 赋值,也可以先定义变量 nextMonth 的类型,然后再赋值,此时就可以省略类型名了。

```
var thisMonth = Month.August
var nextMonth : Month = .September
```

图 9.3 枚举型的赋值

枚举成员具有很好的可读性,常用于 switch 语句中进行条件匹配。如图 9.4 所示,通过 switch 语句判断 nextMonth 变量属于上半年,还是属于下半年,并打印相关信息。

```
switch nextMonth {
case .January, .February, .March, .April, .May, .June :
    print("It belongs first half year")
default : print("It belongs second half year")
}
```

图 9.4　用 switch 语句访问枚举型

9.1.3　原始值

枚举型变量或常量的取值称为枚举值,即枚举型中定义的枚举成员。在 C 语言中,每个枚举值都有一个默认的整数与其对应,通常称这个整数为原始值。

而在 Swift 中,枚举成员的原始值使用起来非常灵活。枚举成员可以有原始值,也可以没有原始值。原始值可以为整型,也可以为字符串型、浮点型或者字符型。如果原始值的取值为整型,则可以从任意一个整数开始自动递增,而不一定从 0 开始。枚举型中所有枚举成员的原始值必须为相同的数据类型。

枚举成员的原始值可以是字符串、字符、整型及浮点型。原始值的类型必须在定义枚举型时显式地定义,具体格式为:

```
enum Name:DataTypeOfRawValue { case valueName=value }
```

在枚举型声明中,每个原始值必须是唯一的,即不能存在两个枚举成员具有相同的原始值。如果某个枚举成员的原始值为整型,而其他枚举成员没有显式的原始值,系统会给它们默认的原始值,并且是递增的。

图 9.5 定义了名为 Car 的枚举型。枚举成员的原始值类型为 Int。该枚举型共有 5 个枚举成员,枚举成员 truck 的原始值显式赋值为 0。其他的枚举成员并没有显式地指明其原始值。这种情况下,系统会自动递增第一个枚举成员的原始值,并给其他没有原始值的枚举成员隐式地赋原

```
enum Car : Int {
    case truck = 0
    case sportsCar
    case SUV
    case MPV
    case limo
}
```

图 9.5　原始值的定义

始值。例如,枚举成员 sportsCar 的原始值为 1,SUV 的原始值为 2,MPV 的原始值为 3,limo 的原始值为 4。

9.1.4　访问原始值

如果要获得一个枚举型变量的原始值,可以通过其 rawValue 属性访问。反之,

也可以通过枚举型的方法根据 rawValue 获得其对应的枚举成员值,如图 9.6 所示。

```
var myCar = Car.SUV
var rawValue = myCar.rawValue
myCar = Car.limo
rawValue = myCar.rawValue

let yourCar = Car(rawValue: 1)
```

图 9.6 原始值的获取

9.1.5 字符串型原始值

类似于整型原始值的自动递增方式产生默认值,当枚举型的原始值为字符串型时,也不需要手动为每个枚举成员赋原始值,系统会根据枚举成员的枚举值直接生成默认的原始值。如图 9.7 所示,将枚举型 Car 的原始值类型修改为字符串型,在枚举成员的声明中并没有定义原始值。在打印枚举型常量 myCar 的原始值时,可以看到系统已经默认将其原始值设置为枚举成员名的字符串了。

```
enum Car: String {
    case truck
    case sportsCar
    case SUV
    case MPV
    case limo
}

let myCar = Car.SUV                    SUV
print("\(myCar.rawValue)")             "SUV\n"
```

图 9.7 字符串型原始值

9.2 关联值

Swift 中的枚举型还可以为每个枚举成员设置一个关联的常量或变量,通常称之为关联值。关联值和枚举成员可以配对存储。枚举成员可以有 0 到多个关联值。每个枚举成员的关联值都可以是一个不同的数据类型。

关联值和原始值不同。关联值是在创建将枚举成员赋值给一个常量或变量时设置的,它可以改变的。而原始值是在定义枚举型的时候预先被设置的值,可以理解为默认值。对于某一个枚举成员,它的原始值是不会改变的。

一个枚举型可以有原始值或者关联值,但不能同时有原始值和关联值。

图 9.8 定义了名为 transportFee 的枚举型。该枚举型有 3 个枚举成员,分别为:byAir 表示坐飞机,byCar 表示开车自驾,byTrain 表示坐火车。每个枚举成员都有不同类型的关联值。通过不同类型的关联值可以描述各种交通方式的不同计费方式。

```
enum transportFee {
    case byAir(Int,Int,Int)
    case byCar(Int,Int)
    case byTrain(Int)
}
```

图 9.8　关联值的定义

9.2.1　关联值的赋值

图 9.9 定义了一个 transportFee 类型的变量,并赋初值。在赋值的时候,不仅需要赋值成员值,还要赋值该成员值对应的关联值。当修改变量 fromShanghaiToBeijing 的成员值后,也要同步修改该成员值对应的关联值。

```
var fromShanghaiToBeijing = transportFee.byTrain(299)      byTrain(299)
fromShanghaiToBeijing = .byAir(800, 230, 50)               byAir(800, 230, 50)
```

图 9.9　关联值的赋值

9.2.2　访问关联值

在 switch 语句中,可以通过值绑定的方式提取枚举成员的关联值。如图 9.10 所示,在 switch 语句中对变量 fromShanghaiToBeijing 进行筛选,通过枚举成员的不同关联值进行匹配。

```
switch fromShanghaiToBeijing {
case .byAir(let ticketFee, let tax, let insurance) :
    print("The sum fee is \(ticketFee+tax+insurance)
    by air")
case .byCar(let fuelFee, let highwayFee) : print("The
    sum fee is \(fuelFee+highwayFee) by car")
case .byTrain(let ticketFee) : print("The sum fee is
    \(ticketFee) by Train")
}
```

图 9.10　提取关联值

如图 9.11 所示，如果一个枚举成员的所有关联值都要被提取为常量或者变量，则用一个 var 或者 let 标注在成员名称前即可。

```
switch fromShanghaiToBeijing {
case let .byAir(ticketFee, tax, insurance) :
    print("The sum fee is
    \(ticketFee+tax+insurance) by air")
case let .byCar(fuelFee, highwayFee) :
    print("The sum fee is
    \(fuelFee+highwayFee) by car")
case let .byTrain(ticketFee) : print("The sum
    fee is \(ticketFee) by Train")
}
```

图 9.11　提取关联值的简写方法

9.2.3　可选型的底层实现

数据类型一节中介绍了 Swift 中特有的一种数据类型"可选型"。那么，可选型的底层实现机制是什么呢？学习枚举型之后，就可以回答这个问题了。实际上，可选型是一个特定的枚举型。该枚举型有两个枚举成员，分别为 none 和 some。none 表示可选型没有值的情况；some 表示可选型中含有值，具体值由该枚举成员的关联值表示。下面通过实例说明。

图 9.12 定义了一个字符串型变量 hobby，用来表示兴趣爱好。有的人有兴趣爱好，有的人则没有，因此 hobby 是可选型的。hobby 可以取值为 Basketball，也可以为 nil。通过 switch 语句判断 hobby 的取值情况。由于 hobby 是由枚举型实现的，所以 hobby 只有两种取值情况，即 .none 和 .some。当 hobby 为 nil 时，对应的枚举值为 .none；当 hobby 中有值时，对应的枚举值为 .some，具体的值为 .some 的关联值，可以通过值绑定访问。

```
var hobby: String?

hobby = "Basketball"
hobby = nil

switch hobby {
case .none:
    print("No hobby")
case .some(let hobbyName):
    print("Hobby is \(hobbyName)")
}
```

图 9.12　关联值的典型实例

在 Swift 中，可选型是一个枚举型的典型实例。为了书写方便，Swift 通过运算符 "?" "!" 以及关键字 "nil" 隐藏了可选型的底层实现机制。实际上，nil 和 .none 是完全等价的。

第 9 章 枚举型

练习题

1. 定义一个枚举型 Direction，用来表示方向。该枚举型有 4 个枚举成员，分别为 East、South、West、North，表示东、南、西、北 4 个方向。定义一个函数 adjacentCity()，要求根据提供的方向返回相应位置的城市名称。该函数有一个参数 direction，为 Direction 类型。函数的返回值为字符串型。调用函数 adjacentCity()，将其赋值给一个常量 northNeighbour，打印该常量。

2. 在第 1 题的基础上为枚举型 Direction 增加原始值，类型为整型，并设置枚举成员 East 的原始值为 0。定义一个常量 currentDirection，赋值为 Direction.North，打印该常量的原始值。

3. 在第 1 题的基础上为枚举型 Direction 增加原始值，类型为字符串型。定义一个常量 currentDirection，赋值为 Direction.North，打印该常量的原始值。

4. 模拟在 ATM 上取款的应用场景。定义一个枚举型 OperationFeedback，有两个枚举成员，分别为 Done 和 Fail。Done 表示正常提款，其关联值为整型，表示取款后的余额。Fail 表示提款失败，其关联值为字符串型，表示提款失败的提示信息。定义一个整型变量 balanceOfATM，表示 ATM 中的余额，赋初始值为 10 000。定义一个函数 withdrawFromATM，参数为整型，返回值为 OperationFeedback 型。将提款数额作为参数传给函数 withdrawFromATM()。该函数负责对提款数额和当前 ATM 中的余额进行比较，当余额不足时，返回提款失败的枚举值。当余额充足时，返回正常提款的枚举值。调用函数 withdrawFromATM()，模拟取款 1288 的情况，并将其返回值赋值给变量 getMoney。根据 getMoney 的取值，打印相应的信息。

本章知识点

1. 介绍了枚举型的定义、枚举型的紧凑写法以及如何使用枚举型。
2. 介绍了枚举型的原始值定义和访问方法，以及字符串型原始值。
3. 介绍了枚举型的关联值定义和赋值方法；关联值的访问方法；可选型的底层实现。

参考代码

第 1 题的参考代码如图 9.13 所示。

```swift
enum Direction {
    case East
    case South
    case West
    case North
}

func adjacentCity(direction: Direction) -> String {
    switch direction {
    case .East:   return "Sanhe"
    case .West:   return "Baoding"
    case .South:  return "Tianjin"
    case .North:  return "Zhangjiakou"
    }
}

let northNeighbour = adjacentCity(direction: .North)
print("The northern city is \(northNeighbour)")
```

图 9.13 第 1 题的参考代码

第 2 题的参考代码如图 9.14 所示。

```swift
enum Direction: Int{
    case East=0
    case South
    case West
    case North
}

let currentDirection = Direction.North
print("The raw value of North is \(currentDirection.rawValue)")
```

图 9.14 第 2 题的参考代码

第 3 题的参考代码如图 9.15 所示。

```swift
enum Direction: String{
    case East
    case South
    case West
    case North
}

let currentDirection = Direction.North
print("The raw value of North is \(currentDirection.rawValue)")
```

图 9.15 第 3 题的参考代码

第 4 题的参考代码如图 9.16 所示。

```swift
enum OperationFeedback {
    case Done(currentBalance: Int)
    case Fail(warningInfo: String)
}

var balanceOfATM = 10000

func withdrawFromATM(amount: Int) -> OperationFeedback{
    if balanceOfATM <= amount {
        balanceOfATM -= amount
        return .Done(currentBalance: balanceOfATM)
    } else {
        return .Fail(warningInfo: "Balance is not enough!")
    }
}

var getMoney = withdrawFromATM(amount: 1288)

switch getMoney {
case .Done(let currentBalance):
    print("Operation is successful. The current balance is \(currentBalance)")
case .Fail(let warningInfo):
    print(warningInfo)
}
```

图 9.16 第 4 题的参考代码

第 10 章 结构体与类

在 Swift 中,结构体和类都是实现面向对象的重要手段,两者类似,又有一定的差异。可以将结构体理解为一种轻量级的类。在 Swift 中,通常通过一个单独的文件定义一个类或者结构体,然后由系统自动生成外部接口。而在 C 语言中,要手动创建接口或者实现文件。因此,在 Swift 中使用类和结构体非常方便。本章将首先从结构体和类的定义介绍两者的相同点,然后再从值类型和引用类型角度讨论两者的差异,最后介绍结构体的使用方法和具体应用场景。

10.1 基本概念

在 Swift 中,结构体拥有很多类的特点:结构体中可以定义属性和方法,使用下标语法、构造器等。但是,结构体并不支持类的继承、析构、引用计数及类型转换等特点。

结构体和类的定义方式非常类似,都是关键字加上名字,然后在花括号中定义具体内容。两者唯一的区别是:结构体通过关键字 struct 标识,而类通过关键字 class 标识。具体的格式如下:

```
struct NameOfStruct {
    //struct definition
}
class NameOfClass {
    //class definition
}
```

结构体和类的命名方面,建议和苹果官方保持一致,即在定义结构体或类的名称时,首字母大写,而在定义结构体或类中的属性和方法时,首字母小写。当名称含有多个单词时,采用驼峰拼写法。代码的命名方法和官方一致,会有更好的可读性。

10.1.1 结构体与类的定义

图 10.1 定义了一个结构体 Book 和一个类 Reader。结构体 Book 含有 3 个属性,分别为 name、price、category,表示书籍的书名、价格以及分类。类 Reader 含有 3 个属性,分别为 name、age、favorite,表示读者的名字、年龄以及喜欢的书籍。其中,属性 favorite 为字符串可选型,表示读者可能有喜欢的书籍,也可能暂时没有。

```
struct Book {
    var name = ""
    var price = 0
    var category = "common"
}

class Reader {
    var name = ""
    var age = 16
    var favorite :String?
}
```

图 10.1 结构体实例

10.1.2 实例初始化

结构体和类都能通过构造器生成实例。因此,最简单的实例化方法就是在结构体名或者类型后面直接带上一对空括号。此时,实例的属性会被初始化为默认值。如果要设定实例中属性的初始值,可以在括号中为属性指定初始值。图 10.2 创建了一个 Reader 类的实例 theReader,并且将其属性值初始化为默认值。另外,还创建了一个 Book 结构体的实例 theBook,并且将其属性赋值为指定的初始值。

10.1.3 点号语法访问属性

结构体和类的实例属性都可以通过点号语法访问,即在实例名后通过点号连接具

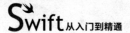

```
let theReader = Reader()
print("Reader's default age is \(theReader.age)")

var theBook = Book(name: "Life of Pi", price: 62, category:
    "adventure")
print("\(theBook.name)'s category is \(theBook.category) and
    price is \(theBook.price)RMB")
```

图 10.2　生成结构体实例

体的属性名即可。如图 10.2 所示，通过点号语法访问结构体实例 theBook 的属性 name、category 以及 price。

10.1.4　点号语法访问方法

结构体中的方法也可以通过点号语法访问。如图 10.3 所示，在结构体 Book 中增加一个方法 description()，用来打印结构体实例的相关信息。然后，创建一个结构体实例 newBook，并用点号语法调用结构体中的方法 description()。

```
struct Book {
    var name = ""
    var price = 0
    var category = "common"
    func description() {
        print("\(name)'s price is \(price),
            category is \(category)")
    }
}

let newBook = Book(name: "Life of Pi", price: 62,
    category: "adventure")
newBook.description()
```

图 10.3　访问结构体中的方法

和结构体一样，类中的方法可以通过点号语法访问。如图 10.4 所示，在类 Reader 中增加一个方法 description()，用来打印类实例的相关信息。然后，创建一个类实例 newReader。在初始化实例 newReader 中的属性时，采用点号语法一一赋初始值，这与结构体不一样。如果要和结构体一样，在创建实例的同时就初始化属性，则需要在类中增加构造器，后面的章节会专门讨论。最后，通过点号语法调用类中的方法 description()。

```
class Reader {
    var name = ""
    var age = 16
    var favorite :String?
    func description() {
        print("\(name) is \(age) year old, favorite
            is \(favorite!)")
    }
}
let newReader = Reader()
newReader.name = "Tommy"
newReader.age = 38
newReader.favorite = "Basketball"
newReader.description()
```

图 10.4 访问类中的方法

10.2 值类型与引用类型

10.2.1 值类型的结构体

结构体是值类型,即当一个结构体变量的值赋给另一个变量时,是通过复制结构体变量值实现的。例如,将结构体变量 temp 赋值给另一个变量 new 时,实际的操作是:根据 temp 所占的存储空间,在内存中重新开辟一片新的存储空间给 new,然后将 temp 的值写入 new 的存储空间中。此时,temp 和 new 对应的存储空间中的值完全相同,但两者之间是完全独立存在的。

在 Swift 中,基本类型包括整型、浮点型、布尔型、字符串、字符型、数组以及字典,都属于值类型,因为它们的底层都是通过结构体实现的。在 Xcode 中,输入 Int 并选中,单击 command 键,在弹出的菜单中选择"Jump to Definition",即可打开 Int 的定义。如图 10.5 所示,整型 Int 是结构体类型。同样,如图 10.6～图 10.8 所示,浮点型、数组、字符串型都是结构体类型。

```
/// A signed integer value type.
///
/// On 32-bit platforms, `Int` is the same size as `Int32`, and
/// on 64-bit platforms, `Int` is the same size as `Int64`.
public struct Int : FixedWidthInteger, SignedInteger {
```

图 10.5 整型的定义

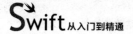

```
/// A double-precision, floating-point value type.
public struct Double {
```

图 10.6　浮点型的定义

```
/// - Note: The `ContiguousArray` and `ArraySlice` types are not bridged;
///   instances of those types always have a contiguous block of memory as
///   their storage.
public struct Array<Element> {
```

图 10.7　数组的定义

```
/// [glossary]: http://www.unicode.org/glossary/
/// [clusters]: http://www.unicode.org/glossary/#extended_grapheme_cluster
/// [scalars]: http://www.unicode.org/glossary/#unicode_scalar_value
/// [equivalence]: http://www.unicode.org/glossary/#canonical_equivalent
public struct String {
```

图 10.8　字符串型的定义

　　如图 10.9 所示，将结构体实例 theBook 的值赋给变量 anotherBook。此时，通过打印输出两个结构体实例的值，可以看到 theBook 和 anotherBook 的属性值完全一致。然后，将实例 anotherBook 中的属性赋值为一个新值，再次打印输出两个结构体实例的值，可以看到 theBook 的值没有变化，而 anotherBook 的值为修改后的新值。这就说明了结构体的值类型特点。结构体实例在进行赋值时，重新复制了一份新的结构体。因此，两个结构体实例在赋值后，虽然值一样，但是其物理存储空间是两个不同的地方。当其中一个结构体实例的值发生改变时，将不会影响到另一个结构体实例的值。

```
var anotherBook = theBook
print(theBook)
print(anotherBook)
anotherBook.category = "history"
anotherBook.price = 136
anotherBook.name = "Empiror Kangxi"
print(theBook)
print(anotherBook)
```

```
Book
"Book(name: "Life of Pi", price: 62, category: "adventure")\n"
"Book(name: "Life of Pi", price: 62, category: "adventure")\n"
Book
Book
Book
"Book(name: "Life of Pi", price: 62, category: "adventure")\n"
"Book(name: "Empiror Kangxi", price: 136, category: "history")\n"
```

图 10.9　值类型的结构体

10.2.2　引用类型的类

　　类是引用类型的。所谓引用类型，就是引用类型变量在被赋值给另一个变量时，是将该引用类型变量的物理存储地址赋值给另一个变量。赋值后,该变量和被赋值的变量指向的物理地址相同。学过 C 语言的读者可以将其理解为指针类型。

如图 10.10 所示,将 Reader 类实例 theReader 赋值给变量 anotherReader。此时,theReader 和 anotherReader 都指向同一片存储 Reader 类实例的空间。通过打印输出这两个实例,可以看到两个实例的属性值完全相同。然后,修改实例 anotherReader 中属性 name 和 age 的值。再次打印这两个实例,发现两个实例的属性值都变成了修改后的值,这就说明了类的引用类型特点。

```
theReader.name = "Tommy"
var anotherReader = theReader
print("\(theReader.name) is \(theReader.age)")
print("\(anotherReader.name) is \(anotherReader.age)")

anotherReader.name = "Jerry"
anotherReader.age = 39
print("\(theReader.name) is \(theReader.age)")
print("\(anotherReader.name) is \(anotherReader.age)")
```

```
Reader
Reader
"Tommy is 16\n"
"Tommy is 16\n"

Reader
Reader
"Jerry is 39\n"
"Jerry is 39\n"
```

图 10.10　引用类型的类

10.2.3　引用类型的等价运算符

引用类型和值类型存在本质差别,因此进行引用类型变量的比较时,不能沿用值类型的比较运算符"=="。运算符"=="表示两个值类型的变量的值相等。在比较引用类型变量时,使用等价运算符"===",即 3 个等号。它表示两个引用类型的变量引用了同一个类实例,如图 10.11 所示。

```
if theReader === anotherReader {
    print("They are equal instances of class.")
} else {
    print("They are not equal.")
}
```

"They are equal instances of class.\n"

图 10.11　引用类型的等价运算符

结构体是值类型,通过值复制传递值。而类为引用类型,通过引用值传递值。因此,结构体和类适用于不同的应用场景。在实际应用中,大部分情况都会使用类描述一个复杂的数据结构或者对象。而结构体则主要用来描述一些相关的简单数据值,如一个几何形状的尺寸或者一个三维坐标系等。

练习题

1. 用结构体定义一个坐标 Coordinate。该结构体中包含两个属性：x 和 y，分别表示 x 坐标和 y 坐标。这两个属性均为浮点型。

2. 用结构体定义一个线段 Line。该结构体中包含两个属性：startPoint 和 endPoint，分别表示起点和终点。另外，结构体还有一个方法 length()，用来计算线段的长度。

3. 定义两个 Coordinate 类型的常量 pointA 和 pointB，分别表示两个点，坐标分别为(1,2)、(3,6)。再定义一个 Line 类型的常量 lineAB，表示由 pointA 和 pointB 组成的线段。打印线段 lineAB 的长度。

4. 用类定义一个洗衣机 WashingMachine。该类中包含 3 个属性：brand、isOn、state，分别表示品牌、开关状态及工作状态。其中，brand 为字符串型，isOn 为布尔型，state 为枚举型 stageOfWashingMachine。该枚举型有 5 个枚举成员，分别表示就绪状态、注水状态、洗涤状态、脱水状态、烘干状态。另外，类中还有一个方法 description()，当洗衣机的开关打开时，打印洗衣机的品牌信息和当前工作状态；否则，提示洗衣机开关关闭了。

5. 定义一个洗衣机类的实例 myWashingMachine，并初始化该实例的 3 个属性。最后，调用实例的方法 description() 打印当前洗衣机的状态信息。

本章知识点

1. 结构体和类的定义以及实例初始化，点号语法的访问属性和方法。
2. 值类型的结构体、引用类型的类、引用类型的等价运算符。

参考代码

第 1～3 题的参考代码如图 10.12 所示。
第 4、5 题的参考代码如图 10.13 所示。

```
struct Coordinate {
    let x : Double
    let y : Double
}

struct Line {
    let startPoint: Coordinate
    let endPoint: Coordinate
    func length() -> Double {
        let x = startPoint.x - endPoint.x
        let y = startPoint.y - endPoint.y
        return sqrt(x*x + y*y)
    }
}

let pointA = Coordinate(x: 1, y: 2)
let pointB = Coordinate(x: 3, y: 6)
let lineAB = Line(startPoint: pointA, endPoint: pointB)

print("The length of line AB is \(lineAB.length())")
```

图 10.12　第 1～3 题的参考代码

```
enum stageOfWashingMachine {
    case ready
    case watering
    case washing
    case dewatering
    case drying
}

class WashingMachine {
    var brand: String = " "
    var isOn: Bool = false
    var state: stageOfWashingMachine = .ready
    func description(){
        if isOn {
            print("Washing machine: \(brand) is \(state)")
        } else {
            print("Washing machine is power off")
        }
    }
}

let myWashingMachine = WashingMachine()
myWashingMachine.brand = "LittleSwan"
myWashingMachine.isOn = true
myWashingMachine.state = .drying
myWashingMachine.description()
```

图 10.13　第 4、5 题的参考代码

第 11 章 属性

本章主要介绍属性相关的概念和知识。前面章节介绍的类、结构体和枚举型都可以有属性。属性将值和这些类型进行了关联。属性分为存储属性、计算属性和类型属性。存储属性通过常量或者变量存储实例的值。计算属性是用来计算值的。存储属性和计算属性与具体的实例相关联。类型属性是属于类型本身的,不属于特定的实例。

另外,本章还将介绍属性观察者,用来监视属性值的变化,并由此触发特定的操作。

11.1 存储属性

存储属性就是存储在一个类或一个结构体中的变量或常量,既可以在定义存储属性的时候赋一个默认值,也可以在构造器中设置或修改存储属性的值。

11.1.1 常量存储属性

图 11.1 定义了一个结构体 User,该结构体包含 3 个存储属性,分别为常量属性 id,变量属性 name、password 和 email。然后,创建了一个结构体 User 的实例 theUser,并通过调用默认构造器的方式初始化 theUser 中的存储属性。注意:常量存储属性被初始化后,值是不能改变的。在本例中,初始化后实例 theUser 的 id 为 16,当将 19 赋值给 id 时,系统报错。

```
struct User {
    let id : Int
    var name : String
    var password : String
    var email : String
}

var theUser = User(id: 16, name: "Tommy", password:
    "bhq963", email: "tommy@gmail.com")

theUser.id = 19    ⊘ Cannot assign to property: 'id' is a 'let' constant
```

图 11.1 类的存储属性

11.1.2 变量存储属性

对实例中的其他变量存储属性进行修改是允许的。如图 11.2 所示，修改了实例 theUser 的变量存储属性 name、password、email。

```
theUser.name = "pennie"
theUser.password = "5263tt"
theUser.email = "pennie@gmail.com"
```

图 11.2 修改类的存储属性

11.1.3 常量实例的存储属性

上例中创建了结构体 User 的实例，并将其赋值给变量 theUser，程序运行正常。如果将实例赋值给一个常量 anotherUser，那么该常量结构体实例中存储属性的值都不能被改变，包括常量存储属性和变量存储属性。如图 11.3 所示，创建了 User 的一个实例并进行了初始化。然后，将其赋值给常量 anotherUser。此后，对常量结构体实例 anotherUser 的常量存储属性 id 和变量存储属性 name、password、email 赋值均会导致系统报错，因为该结构体实例为常量。

```
let anotherUser = User(id: 17, name: "Sammy", password:
    "urs33", email: "sammy@gmail.com")

anotherUser.id = 18       ⊘ Cannot assign to property: 'id' is a 'let' constan
anotherUser.name = "sam"  ⊘ Cannot assign to property: 'anotherUse
anotherUser.password = "newp"  ⊘ Cannot assign to property: 'anot…
anotherUser.email = "new@gmail.com"  ⊘ Cannot assign to propert…
```

图 11.3 常量存储属性

11.1.4 延迟存储属性

在存储属性声明前加上一个关键字 lazy,表示该属性为延迟存储属性。延迟存储属性在实例第一次被调用的时候才会计算初始值。因此,延迟存储属性必须声明为变量。而常量属性在实例初始化完成后必须有值。

当一个属性在初始化时需要占用大量系统资源(时间或空间)时,就可以声明为延迟属性。当一个属性的值依赖于实例初始化完成后的外部数据时,也将其声明为延迟属性。如图 11.4 所示,将结构体 User 修改为类,并对其属性赋初值。然后,在类 User 中继续定义一个延迟属性 image,该属性为结构体 ImageInfo 的一个实例。创建类 User 的实例 theUser 时,初始化了除延迟属性 image 外的所有存储属性,直到第一次在 print 语句中使用到该属性时,才创建了 image 实例。

```
class User {
    var id = 0
    var name = ""
    var password = ""
    var email = ""
    lazy var image = ImageInfo()
}

struct ImageInfo {
    var name = "default name"
    var path = "default path"
}

var theUser = User()
theUser.id = 18
theUser.name = "sammy"
print("the name of image is \(theUser.image.name)")
```

图 11.4 延迟存储属性

11.2 计算属性

存储属性是最常见的一种属性。除了存储属性,还有计算属性,它经过计算后,才能返回属性值。存储属性可以为常量或变量,而计算属性只能为变量。计算属性是 Swift 中特有的一种属性,它不直接存储值,而是通过 getter()和 setter()方法获取和设置值。类、结构体和枚举型都可以定义计算属性。

11.2.1 get()方法

如图 11.5 所示,在上例的基础上,向类 User 的定义中增加一个计算属性 holiday。该属性值的获取需要通过该属性的 get()方法计算得到。假期 holiday 的值由工作年限 workingYear 的值计算而来,即工作年限小于 1 年的,假期为 5 天;工作年限为 1~5 年的,假期为 5 天加上工作年限数;工作年限超过 5 年的,假期均为 12 天。本例创建了一个类 User 的实例 theUser,并设置该实例的 workingYear 为 3,由 get()方法计算可以得到 holiday 为 8 天。

```
struct ImageInfo {
    var name = "default name"
    var path = "default path"
}
class User {
    var id = 0
    var name = ""
    var password = ""
    var email = ""
    lazy var image = ImageInfo()
    var workingYear = 0
    var holiday : Int {
        get {
            var days : Int
            switch workingYear {
                case 0: days = 5
                case 1...5: days = 5 + workingYear
                default: days = 12
            }
            return days
        }
    }
}
var theUser = User()
theUser.name = "Tony"
theUser.workingYear = 3
print("Tony has worked for \(theUser.workingYear) years and he has holiday:
    \(theUser.holiday) days")
```

图 11.5 计算属性的 get()方法

通过实例可以看出,计算属性本身并不存储数据,但可以通过计算返回数据。而在类以外,对计算属性和存储属性的访问方法是一样的。

11.2.2 set()方法

计算属性的 get() 方法可以通过计算读取属性的值。反之，也可以根据计算属性的值设置其他相关存储属性的值。计算属性的 set() 方法就可以实现这一点。注意：计算属性的 set() 方法是没有返回值的。如图 11.6 所示，在计算属性 holiday 中增加一个 set() 方法。该方法根据 newValue 值的大小确定相应的工作年限 workingYear 的计算公式，并对 workingYear 赋值。这里的 newValue 就是 holiday 的新值。每当 holiday 被赋值，都会调用其 set() 方法。

```
var holiday : Int {
    get {
        var days : Int
        switch workingYear {
            case 0: days = 5
            case 1...5: days = 5 + workingYear
            default: days = 12
        }
        return days
    }
    set {
        switch newValue {
            case 5: workingYear = 0
            case 6...11: workingYear = newValue -
                5
            default: workingYear = 6
        }
    }
}
```

图 11.6　计算属性的 set() 方法

如图 11.7 所示，在类 User 的定义外设置实例 theUser 的计算属性 holiday 为 10，此时 holiday 的 set() 方法就会被调用，从而设置 workingYear 的值为 5。通过打印语句可以看到，实例的存储属性 workingYear 已经被设置为 5 了。

```
theUser.holiday = 10
print("Tony has worked for \(theUser.workingYear)
    years.")
```

```
User
"Tony has worked for 5 years.\n"
```

图 11.7　由计算属性设置存储属性

11.3 属性观察器

Swift 提供了一种叫作属性观察器的机制监控属性值的变化。属性观察器也是 Swift 中的一个特色。每当要改变一个属性值之前或之后，都会触发属性观察器。除了延迟存储属性，所有的其他属性都可以增加一个属性观察器，对其值的变化进行监控。

属性观察器有两个方法：willSet()和 didSet()。方法 willSet()在属性的值被改变前触发或被调用。方法 didSet()在属性的值被改变后触发或被调用。方法 willSet()会将新值作为常量参数传入，默认名称为 newValue，也可以自定义参数名称。方法 didSet()则会将旧值作为参数传入 oldValue，同样也可以自定义参数名称。

图 11.8 定义了一个类 Website，包含两个属性 clicks 和 domain。domain 表示域名，初始值为空字符串，clicks 表示单击量，初始值为 0。这里，为 clicks 添加了两个属性观察器：方法 willSet()和方法 didSet()。方法 willSet()中，自定义了参数 newClicks，用来接收新值。当方法 willSet()被调用时，打印单击量 clicks 即将被赋的新值 newClicks。在方法 didSet()中，没有自定义参数接收旧值，默认情况下，旧值将保存在 oldValue 中。当方法 didSet()被调用时，打印单击量的变化情况。最后，定义

```
3   class Website {
4       var domain : String = ""
5       var maxClicks = 10000
6       var clicks : Int = 0 {
7           willSet(newClicks){
8               print("clicks will be set to \(newClicks)")
9           }
10          didSet {
11              print("did set clicks from \(oldValue) to \(clicks)")
12          }
13      }
14  }
15
16  let theWebsite = Website()
17  theWebsite.domain = "www.buaa.edu.cn"
18  theWebsite.clicks = 100
19  theWebsite.clicks = 200
```

```
clicks will be set to 100
did set clicks from 0 to 100
clicks will be set to 200
did set clicks from 100 to 200
```

图 11.8 属性观察器

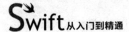

类 Website 的实例 theWebsite，并对属性 clicks 赋了两个不同的值。运行结果显示，在两次赋值的前后，属性观察器都得到正常的触发。

属性观察器可以方便地监控属性值的变化，并能在值的变化前或变化后做出反应。上例中，类 Website 有一个属性 clicks，表示网站允许的单击量。当单击量过大时，可能会导致服务器崩溃。为了防止 clicks 取值过大，可以通过属性观察器 didSet 进行控制。如图 11.9 所示，在类 Website 中添加一个存储属性 maxClicks，用来保存允许的最大单击量。在属性 clicks 观察器 didSet 中增加代码，对变化后的 clicks 值进行判断。如果变化后的 clicks 值大于最大值 maxClicks，则将其回滚到变化前的值 oldValue，并打印提示信息。如果变化后的 clicks 值小于 maxClicks，则打印 clicks 值设置成功的信息。

```
3   class Website {
4       var domain : String = ""
5       var maxClicks = 10000
6       var clicks : Int = 0 {
7           willSet(newClicks){
8               print("clicks will be set to \(newClicks)")
9           }
10          didSet {
11              if clicks > maxClicks {
12                  print("\(clicks) is too high. Fall Back to
                        \(oldValue)")
13                  clicks = oldValue
14              } else {
15                  print("did set clicks from \(oldValue) to
                        \(clicks)")
16              }
17          }
18      }
19  }
20
21  let theWebsite = Website()
22  theWebsite.domain = "www.buaa.edu.cn"
23  theWebsite.clicks = 100
24  theWebsite.clicks = 200
25  theWebsite.clicks = 20000
```

```
clicks will be set to 100
did set clicks from 0 to 100
clicks will be set to 200
did set clicks from 100 to 200
clicks will be set to 20000
20000 is too high. Fall Back to 200
```

图 11.9 属性观察器的应用

11.4 类型属性

每次类型实例化后，每个实例都拥有自己独立的属性值。如果要让某个类型的所有实例都共享同一个属性，就需要引入类型属性的概念。

类型属性是用来表示一个类型的所有实例都共享的属性，如该类型的所有实例都能访问的常量或变量。如果值类型的类型属性也是存储属性，则可以为常量或变量；如果值类型的类型属性也是计算属性，则只能为变量。

类型属性使用关键字 static 标识。类型属性作为类型定义的一部分，它的作用域为类型的内部。与实例属性一样，类型属性的访问也是通过点号运算符进行的。但是，类型属性只能通过类型本身获取和修改，不能通过实例访问。

图 11.10 定义了一个类 Visitor，表示网站访客。类 Visitor 定义了两个存储属性 name 和 stayTime，表示访客的名字和在网站的停留时间。另外，Visitor 还定义了访问权限 permission，既是存储属性，又是类型属性，只有类 Visitor 本身才能获取和修改它的值。本例中，首先创建了一个 Visitor 实例为 theVisitor，该实例有自己的 name 和 stayTime。此时，访问权限 permission 为初始值 visitor，表示所有实例的访问权限都为访客。然后，又创建了一个实例为 anotherVisitor，该实例也有自己的 name 和 stayTime。现在要将所有实例的访问权限都修改为 administrator，即管理员权限，需

```
3   class Visitor {
4       var name : String = ""
5       var stayTime : Int = 0
6       static var permission : String = "visitor"
7   }
8
9   let theVisitor = Visitor()
10  theVisitor.name = "Tom"
11  theVisitor.stayTime = 5
12  print("Current permission is \(Visitor.permission)")
13  let anotherVisitor = Visitor()
14  anotherVisitor.name = "Sam"
15  anotherVisitor.stayTime = 9
16  Visitor.permission = "administrator"
17  print("Now permission is \(Visitor.permission)")
```

```
Current permission is visitor
Now permission is administrator
```

图 11.10　类型属性

要通过类 Visitor 的点号运算修改类型属性 permission 的值。

练习题

1. 定义结构体 Circle,用来表示圆。该结构体含有 3 个属性,分别为圆周率 pi、半径 radius 以及面积 area。其中,area 为计算属性,通过该属性可以得到圆的面积,也可以根据当前面积计算出半径 radius。定义一个结构体 Circle 的实例 smallCircle,并将其半径设为 6。要求打印出实例 smallCircle 的面积。再定义一个结构体 Circle 的实例 bigCircle,并将其面积设为 10 000,要求打印出实例 bigCircle 的半径。

2. 定义一个结构体 Car,用来表示汽车的相关信息。该结构体至少包含 3 个属性:车牌号、品牌、车主姓名。汽车在二手车市场交易时,需要变更车主姓名。变更前,要打印变更提示信息。变更后,要打印新车主和原车主的相关信息。定义一个结构体 Car 的实例 myCar,并对其属性进行初始化,然后再变更车主信息。

3. 定义一个结构体 GamePlayer,表示游戏玩家。该结构体要描述以下信息:游戏所有玩家的当前最高得分、玩家名、玩家 ID、当前所在关卡(共 5 关,第一关通关可得 1000 分、第二关通关可得 9000 分、第三~五关通关均可得 10000 分)、当前关卡所得分、总得分(由所在关卡和当前关卡得分计算而得)。定义两个玩家,分别进行到第二关和第五关,当前关卡得分分别为 876 分和 6690 分。打印两个玩家的各自总得分和游戏中所有玩家的当前最高分。

本章知识点

1. 常量存储属性、变量存储属性、常量实例的存储属性以及延迟存储属性。
2. 计算属性的定义和计算属性的 get() 和 set() 方法。
3. 属性观察器的定义,属性观察器的 willSet() 和 didSet() 方法,以及属性观察器的应用。
4. 类型属性的定义和类型属性的访问。

参考代码

第 1 题的参考代码如图 11.11 所示。

```
struct Circle {
    let pi = 3.1415926
    var radius = 0.0
    var area: Double {
        get {
            return pi * radius * radius      113.0973336
        }
        set {
            radius = sqrt(newValue/pi)       Circle
        }
    }
}

var smallCircle = Circle(radius: 6)          Circle
var area = Int(smallCircle.area)             113
print("Small circle's area is \(area)")      "Small circle's area is 113\n"

var bigCircle = Circle()                     Circle
bigCircle.area = 10000                       Circle
var radius = Int(bigCircle.radius)           56
print("Big Circle's radius is \(radius)")    "Big Circle's radius is 56\n"
```

图 11.11　第 1 题的参考代码

第 2 题的参考代码如图 11.12 所示。

```
struct Car {
    var licenseNumber: String
    var brand: String
    var owner: String {
        willSet {
            if newValue != owner {
                print("Car's owner will be changed.")
            }
        }
        didSet {
            print("Car's ownership has been changed from
                \(oldValue) to \(owner)")
        }
    }
}

var myCar = Car(licenseNumber: "京A05158", brand:
    "Toyota", owner: "Tommy")
myCar.owner = "Sam"
```

图 11.12　第 2 题的参考代码

第 3 题的参考代码如图 11.13 所示。

```swift
struct GamePlayer {
    static var highestScore = 0
    var playerName: String
    var id: Int
    var currentStage: Int
    var currentScore: Int
    var totalScore: Int {
        get {
            switch currentStage {
            case 1: return currentScore
            case 2: return 1000+currentScore
            case 3: return 10000+currentScore
            case 4: return 20000+currentScore
            case 5: return 30000+currentScore
            default: return 0
            }
        }
    }
}

var jack = GamePlayer(playerName: "Jack", id: 9458, currentStage: 2, currentScore: 876)
var tom = GamePlayer(playerName: "Tom", id: 5670, currentStage: 5, currentScore: 6690)
if tom.totalScore < jack.totalScore {
    GamePlayer.highestScore = jack.totalScore
} else {
    GamePlayer.highestScore = tom.totalScore
}
print("Jack's total score is \(jack.totalScore)")
print("Tom's total score is \(tom.totalScore)")
print("Current highest score is \(GamePlayer.highestScore)")
```

图 11.13　第 3 题的参考代码

第 12 章 方 法

方法是在类、结构体或枚举中定义的，用来实现具体任务或功能的函数。方法分为实例方法和类型方法。实例方法与实例关联。类型方法只与类型本身关联，与该类型的实例无关。本章分别介绍实例方法和类型方法的概念和使用。

12.1 实例方法

实例方法是指类、结构体或枚举型等实例中的方法。实例方法可以访问和修改实例的属性，实现特定的功能。实例方法的定义和函数完全一致。

12.1.1 实例方法的定义和调用

实例方法的定义要写在类型定义的花括号内。实例方法可以隐式地访问属于同一个类型的其他实例方法和属性。实例方法只能被实例调用。图 12.1 定义了一个类 Website，该类含有一个属性 visitCount，表示访问次数，以及一个实例方法 visiting()，用来进行访问计数。实例方法 visiting() 没有参数，实现访问次数 visitCount 的自增。定义完类 Website 后，创建一个该类的实例 sina。然后，多次访问实例的属性 visitCount，并调用实例的方法 visiting()。可以看到，每调用一次实例方法 visiting()，实例属性就会增加 1。

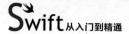

```
class Website {
    var visitCount = 0
    func visiting(){
        visitCount += 1
    }
}

let sina = Website()          Website
sina.visitCount               0
sina.visiting()               Website
sina.visitCount               1
sina.visiting()               Website
sina.visitCount               2
```

图 12.1　实例方法

12.1.2　带参数的方法

方法既可以没有参数，也可以有一个或多个参数。从 Swift 4 开始，去掉了局部参数名和外部参数名的概念，将两者统一为方法的参数名。参数名不仅可以在方法内访问，也可以在外部调用方法时作为提示参数输入的信息出现。如图 12.2 所示，修改上例的类 Website 中的方法 visiting()，为其增加了两个参数 visitor 和 visitDate，分别表示访问者和访问日期。定义完类之后，创建一个 Website 的实例 sina。实例 sina 调用其方法 visiting()。在方法调用的时候，参数名 visitor 和 visitDate 作为提示参数输入的信息出现。

```
class Website {
    var visitCount = 0
    var visitor = [String]()
    var visitDate = ""
    func visiting(visitor: String, visitDate : String){
        visitCount += 1
        self.visitor.append(visitor)
        self.visitDate = visitDate
    }
}

let sina = Website()
sina.visiting(visitor: "Tommy", visitDate: "2016-6-1")
sina.visitCount
sina.visitor
sina.visitDate
```

图 12.2　类中方法的参数

每个实例都有一个隐式的属性 self，表示这个实例本身。在实例的方法中可以通过 self 引用实例自己。图 12.2 中的语句"visitCount += 1"等价于"self.visitCount += 1"。一般情况下，实例方法引用实例方法或属性是不需要显式写上 self 的。但是，当实例方法的参数名与实例属性的名字相同时，根据就近原则，编译器按照实例方法的参数处理。此时，如果要引用实例属性，就必须通过 self 进行区分。这里，方法 visiting() 的第一个参数为 visitor，是字符串型。另外，类中有一个实例属性也叫 visitor，是字符串数字型。为了实现向实例属性 visitor 中不断增加新访客的效果，需要显式地使用 self 区分这两个同名的变量。

12.2 类型方法

类型方法是只能由类型本身调用的方法。在类、结构体和枚举中，声明类型方法是通过在方法的前面加上关键字 static 实现的。类型方法和实例方法一样，都是采用点号语法进行调用。不同之处在于，类型方法是类型本身对该方法的调用，而实例方法只能是实例对该方法的调用。在类型方法中，self 指向类型本身，而不是实例。

如图 12.3 所示，在类 Website 的基础上，将方法 visiting() 声明为类型方法，即在方法前面加上关键字 static。结果，编译器报错，并提示在类型方法内不能使用实例属性。因此，在实例属性 visitCount、visitor、visitDate 前均加上关键字 static，将这些属性声明为类型属性，编译器不再报错。定义完类 Website 后，不能通过实例调用类型

```
class Website {
    static var visitCount = 0
    static var visitor = [String]()
    static var visitDate = ""
    static func visiting(visitor:String, visitDate : String){
        visitCount += 1
        self.visitor.append(visitor)
        self.visitDate = visitDate
    }
}

Website.visiting(visitor:"Tommy", visitDate: "2016-6-1")
Website.visitCount
Website.visitor
Website.visitDate
```

图 12.3 类型方法

方法,只能直接使用类 Website 调用类型方法 visiting(),并用点号语法调用类型属性。系统编译通过,运行结果正确。

12.3 可变方法

在类 Website 中,可以通过实例方法或者类型方法修改属性的值。例如,图 12.2 中,通过实例方法 visiting() 修改属性 visitCount 的值;图 12.3 中,通过类型方法 visiting() 修改属性 visitCount、visitor、visitDate 的值。

如果将 Website 改写为结构体,则编译器会报错,提示结构体的方法不能修改属性,如图 12.4 所示。

```
struct Website {
    var visitCount = 0
    func visiting() {
        visitCount += 1
```
> Left side of mutating operator isn't mutable: 'self' is immutable
> Mark method 'mutating' to make 'self' mutable   Fix

图 12.4 结构体中的方法修改属性值

如果要使结构体中的方法能够修改属性值,就必须在方法名前面加上关键字 mutating,如图 12.5 所示。

```
struct Website {
    var visitCount = 0
    mutating func visiting() {
        visitCount += 1
    }
}
var sohu = Website()          Website
sohu.visiting()               Website
print("\(sohu.visitCount)")   "1\n"
```

图 12.5 结构体中的可变方法

12.4 下标方法

下标是一种通过下标索引获取值的快捷方法。最典型的例子是在数组中使用下标读写数组元素,如 Array[Index]。在类、结构体和枚举型中,可以自定义下标方法,从而实现对实例属性的赋值和访问。自定义的下标可以有多种索引值类型，从而实现按照不同索引对实例属性的赋值和访问。

下标是通过向实例后的方括号中传入一个或者多个索引值对实例中的属性进行访问和赋值的。自定义一个下标方法要使用关键字 subscript，并显式地声明一个或多个传入的参数类型和返回值类型。在自定义下标方法中，通过 set() 和 get() 方法的定义实现读写或者只读。

具体格式为：

```
subscript(index: Int)->Int {
    get {
        //return index
    }
    set(newValue) {
        //set new value
    }
}
```

这里，newValue 的类型必须和下标方法定义的返回类型一致。如果没有显示地声明 newValue，系统也会提供一个默认的 newValue。

如图 12.6 所示，在类 Website 的基础上增加下标的定义。下标方法的传入参数为整型的 index，返回值为字符串类型。下标方法中定义了 get() 方法和 set() 方法。get() 方法根据索引值 index 返回 visitor 数组中的某一个访客的名字，而 set() 方法则实现向索引值对应的变量进行赋值。在定义完类后，创建了 Website 的实例 sina，并通过方法 visiting() 向数组 visitor 中写入了两个访客的名字。由于定义了下标方法，就可以通过下标语法快捷地对特定元素进行访问和赋值了。这里通过 sina[0] 将 visitor 数组中的第一个值读出，然后通过 sina[1]="Pennie" 实现了向 visitor 数组中特定位置写入值。

下标提供了一种访问集合、列表或序列中元素的快捷方式。在类或结构体中，也可以自定义下标方法实现特定的快捷访问功能。下标的具体含义取决于特定的应用场景。

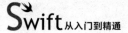

```
class Website {
    var visitCount = 0
    var visitor = [String]()
    var visitDate = ""
    func visiting(visitor:String, visitDate : String){
        visitCount += 1                                     (2 times)
        self.visitor.append(visitor)                        (2 times)
        self.visitDate = visitDate
    }
    subscript(index : Int) -> String {
        get {
            return visitor[index]                           (2 times)
        }
        set {
            visitor[index] = newValue                       "Pennie"
        }
    }
}

var sina = Website()                                        Website
sina.visiting(visitor: "Tom", visitDate: "2016-6-3")        Website
sina.visiting(visitor: "Sam", visitDate: "2016-6-9")        Website
print("\(sina[0])")                                         "Tom\n"
sina[1] = "Pennie"                                          "Pennie"
print("\(sina[1])")                                         "Pennie\n"
```

图 12.6 定义下标

练习题

1. 定义一个结构体，表示篮球运动员。该结构体包含 4 个属性：球员名字、3 分球投中次数、2 分球投中次数以及得分。其中，得分是根据 3 分球投中次数和 2 分球投中次数计算出来的，因此得分为计算属性。

2. 定义一个类，表示篮球队。该类包含 3 个属性：队名、球员、球队得分。球员为第 1 题中定义的篮球运动员类型的数组。球队得分由队中所有球员的得分相加得到，因此球队得分也是计算属性。另外，该类有一个初始化方法，该方法有一个参数，用来表示球队名。该类还有一个向球队增加球员的方法。

3. 在前两题的基础上，至少定义 3 个篮球球员实例，并初始化其姓名、3 分球投中次数以及 2 分球投中次数。再定义一个篮球队，并初始化篮球队名。将这 3 个篮球球员加入到篮球队中，并打印输出该篮球队的得分以及每个球员的得分明细。

4. 定义会计工具包 ToolSet 结构体。该结构体中只有一个类型方法 taxPayment()，用于根据工资额计算应缴纳的税款。该方法有一个参数 salary，浮点型，返回值也是浮点型。缴税额的简易计算方法：工资低于 3000 元时，按照 3% 缴纳；工资低于

12 000 元时，3000 元及 3000 元以上的部分按照 10% 缴纳；工资低于 25 000 元时，12 000元及 12 000 元以上的部分按照 20% 缴纳；工资大于或等于 25 000 元时，25 000 元以上的部分按照 30% 缴纳。调用该类型方法，计算当工资为 25 000 元时应缴纳的税款。

5. 修改第 1 题中的结构体"篮球运动员"，向其中增加两个方法。第一个方法模拟投篮动作，当投中一个 2 分球时，2 分球投中次数加 1；当投中一个 3 分球时，3 分球投中次数加 1。第二个方法负责将该篮球运动员实例的相关信息打印出来，包括球员名字、2 分球投中次数、3 分球投中次数以及总得分。

6. 在第 2 题的类"篮球队"的定义中增加一个下标方法，从而可以通过篮球队的实例的下标访问和设置其中某一个篮球运动员。定义一个篮球队的实例，并通过下标对其中的篮球运动员进行访问和设置。请问能否通过实例下标方法向篮球队实例中添加新的运动员？说明原因。

本章知识点

1. 实例方法的定义和调用，以及带参数的方法。
2. 类型方法的定义和调用。
3. 可变方法的定义。
4. 下标的定义和使用。

参考代码

第 1 题的参考代码如图 12.7 所示。

```
struct BasketballPlayer {
    var name: String
    var threePointShot: Int
    var twoPointShot: Int
    var score: Int {
        get {
            return 2*twoPointShot+3*threePointShot
        }
    }
}
```

图 12.7　第 1 题的参考代码

第 2 题的参考代码如图 12.8 所示。

```swift
class BasketballTeam {
    var name: String
    var players: [BasketballPlayer]
    var score: Int {
        get {
            var sum = 0
            for player in players {
                sum += player.score
            }
            return sum
        }
        
    }
    func addNewPlayer(newPlayer: BasketballPlayer) {
        players.append(newPlayer)
    }
    init(name: String) {
        self.name = name
        players = []
    }
}
```

图 12.8　第 2 题的参考代码

第 3 题的参考代码如图 12.9 所示。

```swift
var player1 = BasketballPlayer(name: "YaoMing", threePointShot: 1, twoPointShot: 15)
var player2 = BasketballPlayer(name: "McGrady", threePointShot: 9, twoPointShot: 12)
var player3 = BasketballPlayer(name: "Alston", threePointShot: 3, twoPointShot: 3)

var houstonRocket = BasketballTeam(name: "Houston Rocket")
houstonRocket.addNewPlayer(newPlayer: player1)
houstonRocket.addNewPlayer(newPlayer: player2)
houstonRocket.addNewPlayer(newPlayer: player3)
print("Total score of \(houstonRocket.name) is \(houstonRocket.score)")
print("Each player's score is:")
for player in houstonRocket.players {
    let name = player.name
    let score = player.score
    print("\(name) score is \(score)")
}
```

图 12.9　第 3 题的参考代码

第 4 题的参考代码如图 12.10 所示。

第 5 题的参考代码如图 12.11 所示。

```swift
struct ToolSet {
    static func taxPayment(salary: Double) -> Double {
        var tax = 0.0
        if salary < 0.0 {
            print("Salary is negtive!")
        }
        if salary < 3000 {
            tax = salary * 0.03
        }else if salary < 12000 {
            tax = 900 + (salary-3000) * 0.1
        }else if salary < 25000 {
            tax = 900 + 900 + (salary - 12000) * 0.2
        } else if salary >= 25000{
            tax = 900 + 900 + 2600 + (salary-25000) * 0.3
        }
        return tax
    }
}

ToolSet.taxPayment(salary: 25000)
```

图 12.10　第 4 题的参考代码

```swift
 3  struct BasketballPlayer {
 4      var name: String
 5      var threePointShot: Int
 6      var twoPointShot: Int
 7      var score: Int {
 8          get {
 9              return 2*twoPointShot+3*threePointShot
10          }
11      }
12      mutating func shot(point: Int) {
13          if point == 3 {
14              threePointShot += 1
15          } else if point == 2 {
16              twoPointShot += 1
17          }
18      }
19      func description() {
20          print("Player \(name) has \(threePointShot) three point shot and
                \(twoPointShot) two point shot, the total socre is \(score)")
21      }
22  }
23
24  var player = BasketballPlayer(name: "YaoMing", threePointShot: 0,
        twoPointShot: 0)
```

图 12.11　第 5 题的参考代码

```
25     player.name = "YaoMing"
26     player.description()
27     player.shot(point: 2)
28     player.shot(point: 2)
29     player.shot(point: 3)
30     player.shot(point: 2)
31     player.shot(point: 2)
32     player.description()
```

```
Player YaoMing has 0 three point shot and 0 two point shot, the total socre is 0
Player YaoMing has 1 three point shot and 4 two point shot, the total socre is 11
```

图 12.11　第 5 题的参考代码(续)

第 6 题的参考代码如图 12.12 所示。

```swift
class BasketballTeam {
    var name: String
    var players: [BasketballPlayer]
    var score: Int {
        get {
            var sum = 0
            for player in players {
                sum += player.score
            }
            return sum
        }
    }
    func addNewPlayer(newPlayer: BasketballPlayer) {
        players.append(newPlayer)
    }
    init(name: String) {
        self.name = name
        players = []
    }
    subscript(index: Int) -> BasketballPlayer {
        get {
            return players[index]
        }
        set {
            players[index] = newValue
        }
    }
}

var player1 = BasketballPlayer(name: "YaoMing", threePointShot: 1, twoPointShot: 15)
var player2 = BasketballPlayer(name: "McGrady", threePointShot: 9, twoPointShot: 12)
var player3 = BasketballPlayer(name: "Alston", threePointShot: 3, twoPointShot: 3)
```

图 12.12　第 6 题的参考代码

```
var houstonRocket = BasketballTeam(name: "Houston Rocket")
houstonRocket.addNewPlayer(newPlayer: player1)
houstonRocket.addNewPlayer(newPlayer: player2)
houstonRocket.addNewPlayer(newPlayer: player3)

houstonRocket[0] = player3
houstonRocket[2] = player1

print("Total score of \(houstonRocket.name) is \(houstonRocket.score)")
print("Each player's score is:")
for i in 0...(houstonRocket.players.count-1) {
    let name = houstonRocket[i].name
    let score = houstonRocket[i].score
    print("\(name) score is \(score)")
}
```

图 12.12 （续）

第 13 章
类

前面的章节介绍了结构体和类的基本概念和用法,以及两者的异同点。本章将进一步讨论类所独有的一些特性,包括类的继承性、类的重载、类的构造以及类的析构。

13.1 继承性

继承性是指一个类可以继承其他类的方法和属性。这个继承的类称为子类,被继承的类称为父类。继承性是类的基本特征。在 Swift 中,一个类可以调用父类的方法,访问父类的属性和下标,还可以重载父类的方法、属性和下标。

13.1.1 基类

没有父类的类称为基类。在 Swift 中,所有的类并不是从一个通用的基类继承而来的。如果不为一个类指定父类,那么这个类本身就是基类。

图 13.1 定义了一个基类 Student。这个类中包括 3 个存储属性 name、age、id,分别表示姓名、年龄和学号。另外,该类还有一个计算属性 basicInfo,用来返回学生的基本信息。basicInfo 的值由 name、age、id 计算而来。该类中有 chooseClass() 和 haveClass() 两个方法,分别实现选课和上课的功能。最后,创建了一个 Student 的实例 theStudent,并为其实例属性赋初始值,然后输出计算属性 basicInfo。这里的类 Student 就是基类。从本例中可以看出,基类是用来定义具体子类的一个基础,它规定了子类最基本的信息和功能。

```
class Student {
    var name = ""
    var age = 0
    var id = ""
    var basicInfo : String {
        return "\(name) is \(age) years old, the id is \(id)"
    }
    func chooseClass(){
        print("\(name) choose a class.")
    }
    func haveClass(){
        print("\(name) have a class.")
    }
}

let theStudent = Student()
theStudent.name = "Tommy"
theStudent.age = 19
theStudent.id = "37060115"
print(theStudent.basicInfo)
```

图 13.1 定义基类

13.1.2 子类

子类就是在一个已有的类的基础上创建的一个新类。子类继承了父类的全部特性，并且还有自己独有的特性，即独有的属性或方法。

定义一个子类的格式为：

```
class SonClass: FatherClass {
    definition
}
```

定义子类时，需要在子类的类名后标注其父类的类名，并用冒号分隔，其他和一般类的定义一样。

如图 13.2 所示，在基类 Student 的基础上，定义了研究生类 Graduate。这里，Student 类和 Graduate 类之间是父子关系。Graduate 类继承了 Student 类中所有的属性和方法。另外，Graduate 类还定义了只属于它自己的存储属性 supervisor 和 researchTopic，用来表示研究生的导师和研究方向。Graduate 类还有自己特有的方法 chooseSuperVisor，用来实现研究生选导师的功能。定义完类以后，创建了一个 Graduate 类的实例 theGraduate。可以看到，实例 theGraduate 不仅可以访问

Graduate 类中的属性和方法，还能访问 Graduate 的父类 Student 中的属性 name、age、id，也能调用父类 Student 中的方法 haveClass()。

```
class Graduate : Student {
    var supervisor = ""
    var researchTopic = ""
    func chooseSuperVisor(superVisor:String){
        self.supervisor = superVisor
    }
}

let theGraduate = Graduate()
theGraduate.name = "Sam"
theGraduate.age = 23
theGraduate.id = "SY0602115"
theGraduate.haveClass()
theGraduate.researchTopic = "Graphics"
theGraduate.chooseSuperVisor(superVisor: "Ian")
print("Graduate \(theGraduate.name) is \(theGraduate.age) and
    the id is \(theGraduate.id), The research topic is
    \(theGraduate.researchTopic) and supervisor is
    \(theGraduate.supervisor)")
```

图 13.2 定义子类

子类也可以有子类。子类的子类也继承父类及父类的父类的全部属性和方法。前面例子中，Graduate 类的父类为 Student 类，Student 类是基类。如图 13.3 所示，以 Graduate 类为父类，定义了它的子类：博士生类 Doctor。Doctor 类不仅继承其父类 Graduate 类的所有属性和方法，同时还继承基类 Student 的所有属性和方法。在 Doctor 类中定义了存储属性：已发表文章 articles，用字符串数组 articles 表示。另外，还定义了一个方法：发表文章 publishArticle，负责将新发表的文章添加到文章列表中。

定义完类以后，创建了 Doctor 类的实例 theDoctor。由运行结果可以看出，实例 theDoctor 不仅可以操作基类的属性 name、age、id 和调用基类的方法 basicInfo()，而且还可以操作其父类的属性 supervisor、调用父类的方法 chooseSuperVisor()。另外，theDoctor 实例还调用了自己独有的方法 publishArticle()，并访问了自己独有的属性 articles 数组。

通过上面的例子不难看出，子类通过继承父类的属性和方法，避免了大量重复性代码。子类在继承父类的时候，要遵守以下两条规则。

```
class Doctor: Graduate {
    var articles = [String]()
    func publishArticle(article : String){
        articles.append(article)
    }
}

let theDoctor = Doctor()
theDoctor.name = "Pennie"
theDoctor.age = 26
theDoctor.id = "BY0607120"
theDoctor.basicInfo
theDoctor.chooseSuperVisor(superVisor: "Ellis")
theDoctor.supervisor
theDoctor.publishArticle(article: "Petri nets theory")
theDoctor.publishArticle(article: "Process management")
theDoctor.articles
```

图 13.3 子类的实例

（1）子类只能有一个父类，即单继承。例如，Graduate 类的父类是 Student，那么就不能再定义别的类为 Graduate 的父类了。

（2）类的继承深度没有限制。例如，Student 类的子类为 Graduate 类，而 Graduate 类的子类为 Doctor 类，这里的父子关系的深度就是 3 层。如果深度没有限制，就意味着还可以继续向下发展子类。

13.1.3　子类的多态性

多态性就是编译器根据上下文判断一个子类实例应该表现为哪一个类（父类，还是子类）的实例。多态性是继承性的一个体现。如图 13.4 所示，在类 Student 和类 Graduate 的定义基础上，另外定义一个函数 sportsGameRoster()。该函数有一个参数 student 为 Student 类，返回值为字符串型，负责将参加运动会的运动员信息以字符串的形式返回。然后定义了一个 Student 类的实例 studentTom，并进行了初始化。另外，还定义了一个 Graduate 类的实例 graduateJim，并进行了初始化。Graduate 类为 Student 类的子类，因此，Graduate 类的实例 graduateJim 既可以表现为 Graduate 类的实例，也可以表现为 Student 类的实例。编译器根据上下文判断实例 graduateJim 应该当成哪个类使用。本例将 graduateJim 实例作为参数传入函数

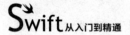

sportsGameRoster()中，根据函数的定义，传入参数为 Student 类型，因此这里 graduateJim 实例就被编译器当成 Student 类的实例传入函数中。

```
func sportsGameRoster(student: Student) -> String {
    return "Athlete name: \(student.name), age: \(student.age), id:
        \(student.id)"
}

let studentTom = Student()
studentTom.name = "Tom"; studentTom.age = 19; studentTom.id =
    "37060116"
let graduateJim = Graduate()
graduateJim.name = "Jim"; graduateJim.age = 24; graduateJim.id =
    "SY060218"

let rosterTom = sportsGameRoster(student: studentTom)
```
Athlete name: Tom, age: 19, id: 37060116
```
let rosterJim = sportsGameRoster(student: graduateJim)
```
Athlete name: Jim, age: 24, id: SY060218

图 13.4　多态性实例

练习题

1. 定义一个交通工具类 Vehicle。该类有如下属性，分别为用途 use、时速 speed、生产商 producer、使用时间 usedYear 以及基本信息 basicInfo（由其他属性的值拼接而成）。另外，还有一个方法 description()，用于打印 Vehicle 的相关信息。定义一个该类的常量实例 train，初始化该实例的各个属性，并调用实例方法 description()。

2. 定义 Vehicle 的子类 Car。该类有两个属性，分别为车主 owner 和车牌照 license。另外，还有一个方法 propertyTransfer()，用来变更车主。定义一个该类的常量实例 myCar，并初始化该实例的各个属性，包括 Car 中定义的属性及其父类的属性。调用实例的方法 description() 和 propertyTransfer()。

3. 定义一个函数 annualInspection()，用来提示一个交通工具是否需要年检。如果需要年检，则返回 true。该函数根据交通工具的使用时间决定是否要年检。如果大于 5，则需要年检，否则不需要年检。分别以实例 train 和 myCar 作为参数调用函数 annualInspection()。

13.2 重载

重载是面向对象的一个非常重要的概念。重载是指子类把从父类继承来的实例方法、类方法、实例属性及下标等，提供一套自己的实现。当子类重载父类的特性时，需要在该特性前面加上关键字 override。对于编译器来说，关键字 override 意味着要重新定义一个特性，且该特性是从父类继承来的。如果不加关键字 override，当子类中出现与父类同名的特性时，编译器会报错。

在子类中重载父类的方法、属性或下标时，可以通过前缀 super 语法引入父类的实现，从而使子类的代码可以专注于实现新增的功能。

一般来说，重载父类的方法 method() 时，先通过 super.method 访问父类的 method() 方法，然后再定义子类中新增的实现。重载父类中属性 property 的 getter() 和 setter() 方法时，通过 super.property 访问父类的 property 属性。重载父类中的下标时，通过 super[index] 访问父类中的相同下标。

13.2.1 重载方法

在子类中，通过提供一个新的实现重载父类的实例方法或类方法。重载后，父类中的方法将会被完全覆盖。如图 13.5 所示，子类 Graduate 重载了父类的方法 chooseClass()。当通过类 Graduate 的实例 theGraduate 调用方法 chooseClass() 时，执行的是子类 Graduate 中的方法，而不是父类 Student 中的同名方法，因为父类的方法 chooseClass() 在子类中被重载后被完全覆盖了。

```
class Graduate : Student {
    var supervisor = ""
    var researchTopic = ""
    func chooseSuperVisor(superVisor:String){
        self.supervisor = superVisor
    }
    override func chooseClass() {
        print("graduate \(name) choose a class")
    }
}
let theGraduate = Graduate()
theGraduate.name = "Sam"
theGraduate.chooseClass()
```

图 13.5 重载方法

13.2.2 重载属性

在子类中也可以重载父类的属性,包括实例属性和类属性,为这些属性提供新的 getter()和 setter()方法,以及添加属性观察器。

重载父类的属性时,子类需要把它的名字和类型都写出来。编译器会以此匹配父类中同名同类的属性。如果将父类中的一个只读属性重载为读写属性,只需在子类中为该属性提供 getter()和 setter()方法。但是,不能将父类的一个读写属性重载为一个只读属性。也就是说,如果在重载父类属性时提供了 setter,就一定要提供 getter。如果不想在子类重载属性 property 的 getter 中修改属性值,可以直接通过 super.property 返回属性值。

图 13.6 修改了基类 Student 的子类 Graduate 的定义。Graduate 子类重载了父类 Student 中的计算属性 basicInfo。在重载的属性 basicInfo 中重写了 basicInfo 的 getter()方法。在 getter()方法中,通过 super.basicInfo 调用父类 Student 中的 getter()方法,从而返回父类的 basicInfo 内容,然后再在此基础上拼接新的信息,即导师和研究方向。

```
class Graduate : Student {
    var supervisor = ""
    var researchTopic = ""

    override var basicInfo: String{
        return super.basicInfo + ", supervisor is \(supervisor),
            research topic is \(researchTopic)"
    }
    func chooseSuperVisor(superVisor:String){
        self.supervisor = superVisor
    }
    override func chooseClass() {
        print("graduate \(name) choose a class")
    }
}
let theGraduate = Graduate()
theGraduate.name = "Sam"
theGraduate.age = 25
theGraduate.id = "SY06011125"
theGraduate.supervisor = "Ian"
theGraduate.researchTopic = "Petri Net"
print(theGraduate.basicInfo)
```

图 13.6 重载属性

完成子类 Graduate 的修改后，就创建了类 Graduate 的实例，并对其属性赋值，然后打印 basicInfo 内容。可以看到，实例 theGraduate 的 basicInfo 不仅包含父类 Student 中的 basicInfo 的内容，还包含新增的导师和研究方向的内容。

13.2.3　重载属性中增加观察器

在子类中重载一个属性的时候，可以为该属性增加属性观察器，监控该属性值的变化。如图 13.7 所示，在类 Student 的子类 theGraduate 中重载存储属性 age，并为其增加观察器 didSet 和 willSet。当属性 age 发生变化时，会触发观察器打印出相应的信息。将实例 theGraduate 的属性 age 设置为 25 时，触发了 age 的属性观察器 willSet，从而打印信息"original age will be set to 25"。当对 age 的赋值完成后，又触发了另一个属性观察器 didSet，并打印信息"age is set from 0 to 25"。

```
18  class Graduate : Student {
19      var supervisor = ""
20      var researchTopic = ""
21      override var age : Int {
22          didSet {
23              print("age is set from \(oldValue) to \(age)")
24          }
25          willSet {
26              print("original age will be set to \(newValue)")
27          }
28      }
29      override var basicInfo: String{
30          return super.basicInfo + ", supervisor is \(supervisor),
                research topic is \(researchTopic)"
31      }
32      func chooseSuperVisor(superVisor:String){
33          self.supervisor = superVisor
34      }
35      override func chooseClass() {
36          print("graduate \(name) choose a class")
37      }
38  }
39  let theGraduate = Graduate()
40  theGraduate.name = "Sam"
41  theGraduate.age = 25
```

```
original age will be set to 25
age is set from 0 to 25
```

图 13.7　重载属性中增加观察器

需要注意的是,不能为继承来的常量存储属性或只读计算型属性增加属性观察器,因为常量属性的值是不允许被修改的。

13.2.4 禁止重载

前面介绍了子类可以重载父类的特性。如果要禁止父类的特性被子类重载,就要使用关键字 final 标识出不能被子类重载的特性。父类的特性包括方法、属性以及下标,都可以通过在前面加上关键字 final 禁止子类重载。如果子类重载了父类中标识为 final 的特性,编译器会提示错误。如果要禁止一个类被继承,也可以在类定义前加 final 关键字禁止该类被继承。

练习题

1. 在 13.1 节练习题中,Car 类的实例调用方法 description() 时,只能打印其父类中相关属性的值,无法打印 Car 类中新定义的属性值。通过重载方法 description() 的方式解决这个问题。

2. 同样,访问 Car 类的属性 basicInfo 时,也只有父类 Vehicle 中的相关属性,无法体现 Car 类中新定义的属性值。通过重载属性 basicInfo 的方法解决这个问题。

3. 当 Car 类的实例属性 usedYear 发生变化时,需要在变化发生前和变化发生后分别打印提示信息,内容包括变化前的值和变化后的值。

13.3 类的构造

类的构造就是在使用类、结构体或枚举型的实例前,先通过构造器设置每个存储型属性的初始值,同时执行其他必要的设置初始化工作。构造器就是用来初始化类、结构体或枚举型的。在 Swift 中,构造器不需要返回值,它的任务是为新实例的第一次使用做好准备。

在创建类和结构体的实例时,所有的存储型属性都应该有适当的初始值。在前面的章节中,通过在定义属性的同时为其赋默认值的方式进行实例属性的初始化。本章介绍通过构造器为存储属性初始化的方法。注意:在进行属性初始化的时候,是不会触发任何属性观察器的。

13.3.1 定义构造器

创建一个类型的实例时,构造器会被自动调用。默认的构造器格式为

```
init() {
    //initialization process
}
```

构造器用关键字 init 命名方法,可以没有参数。

如图 13.8 所示,仍以基类 Student 为例,为其增加一个最简单的默认构造器 init,去掉原来的直接在存储属性定义中赋值的部分。创建一个 Student 的实例 theStudent,不需要传入任何参数。从系统执行结果可以看出,默认构造器 init 对实例中的存储属性进行了正确的初始化。

```
class Student {
    var name : String
    var age : Int
    var id : String
    var basicInfo : String {
        return "\(name) is \(age) years old, the id is \(id)"
    }
    init(){
        name = "no name"
        age = 16
        id = "no id"
    }
}

var theStudent = Student()
```

```
name "no name"
age 16
id "no id"
```

图 13.8 基类的默认构造器

13.3.2 自定义构造器

除了前面介绍的无参数默认构造器外,还可以自定义构造器及其参数,包括参数的类型和名字。如图 13.9 所示,在上例的基础上增加两个自定义构造器。这两个自定义构造器的区别在于参数列表不同:第一个自定义构造器有两个参数,分别为

name 和 age；第二个自定义构造器有 3 个参数，分别为 name、age 和 id。注意：虽然第一个自定义构造器只有两个参数，但是在构造器中仍然需要对全部的 3 个存储属性进行初始化，只不过属性 id 的初始化值是由构造器默认给出的，而不需要实例构造器通过参数的方式传入。

```swift
class Student {
    var name : String
    var age : Int
    var id : String
    var basicInfo : String {
        return "\(name) is \(age) years old, the id is \(id)"
    }
    init(){
        name = "no name"
        age = 16
        id = "no id"
    }
    init(name : String, age : Int){
        self.name = name
        self.age = age
        self.id = "no id"
    }
    init(name : String, age : Int, id : String){
        self.name = name
        self.age = age
        self.id = id
    }
}
```

图 13.9　自定义构造器

如图 13.10 所示，定义完自定义构造器后，创建两个实例：theStudent 和 anotherStudent。实例 theStudent 在创建的时候提供了两个参数的值：name 为 "Tom"、age 为 25。初

```swift
var theStudent = Student(name: "Tom", age: 25)
    name "Tom"
    age 25
    id "no id"

var anotherStudent = Student(name: "Sam", age: 29, id: "BY0602115")
    name "Sam"
    age 29
    id "BY0602115"
```

图 13.10　自定义构造器的实例

始化实例的时候,根据实例的参数列表自动匹配并调用第一个自定义构造器。实例 anotherStudent 在创建的时候,提供了 3 个参数值:name 为"Sam"、age 为 29、id 为 "BY0602115"。初始化实例的时候,根据实例的参数列表自动匹配并调用第二个自定义构造器。

13.3.3 无外部参数名的自定义构造器

定义构造器时,可以使用下画线"_"表示该参数的外部名,从而在调用构造器的时候可以省略参数名。图 13.11 用下画线的方式标注了构造器 init 的外部参数名,在创建实例的时候就不再需要提供外部参数名了。

```
class Student {
    var name : String
    var age : Int
    var id : String
    var basicInfo : String {
        return "\(name) is \(age) years old, the id is \(id)"
    }
    init(_ name : String, _ age : Int, _ id : String){
        self.name = name
        self.age = age
        self.id = id
    }
}
var anotherStudent = Student("Sam", 29, "BY0602115")
    name "Sam"
    age 29
    id "BY0602115"
```

图 13.11　下画线标注构造器外部参数名

省略外部参数名可以简化创建实例的语句,但会带来代码可读性的下降,建议读者谨慎使用这种省略外部参数名的构造器。

13.3.4 可选型属性的自动初始化

在类的定义中,如果属性的值有可能为空,必须声明为可选型。对于可选型的属性,系统会自动为该属性初始化为 nil。该属性不需要在构造器里进行初始化。如图 13.12 所示,将 Student 类中的存储属性 id 改为字符串可选型,可以去掉构造器中的 id 初始化语句。创建实例 theStudent 时,系统会自动将其初始化为 nil。

```
class Student {
    var name : String
    var age : Int
    var id : String?
    var basicInfo : String {
        return "\(name) is \(age) years old, the id is \(String(describing: id))"
    }
    init(){
        name = "no name"
        age = 16
    }
    init(name : String, age : Int){
        self.name = name
        self.age = age
    }
}

var theStudent = Student(name: "Tom", age: 25)
    name "Tom"
    age 25
    nil
```

图 13.12 可选型属性的自动初始化

13.3.5 常量属性的初始化

在构造器中可以修改常量属性的值。一旦初始化完成后，常量属性的值就不能被改变了。如图 13.13 所示，将 Student 类中定义的存储属性 id 修改为常量。在 Student

```
class Student {
    var name : String
    var age : Int
    let id : String
    var basicInfo : String {
        return "\(name) is \(age) years old, the id is \(id)"
    }
    init(){
        name = "no name"
        age = 16
        id = ""
    }
    init(name : String, age : Int, id : String){
        self.name = name
        self.age = age
        self.id = id
    }
}

var theStudent = Student(name: "Tom", age: 25, id: "BY0602115")
theStudent.id = "BY0602116"    ⊘ Cannot assign to property: 'id' is a 'let' constant
```

图 13.13 类中的常量属性

类实例化后,再对常量属性 id 进行赋值,编译器会报错,提醒"不能对一个常量属性赋值"。

13.3.6 构造器代理

在定义复杂构造器的时候,可以通过调用已定义的构造器完成部分构造工作,减少重复代码,这种类型的构造器称为构造器代理。值类型和引用类型的构造器代理的实现方式有所不同。

13.3.7 值类型的构造器代理

结构体类型是值类型。值类型的构造器代理只能用于类型本身的构造器。在自定义的构造器中,通过 self.init 调用类型中已经定义的构造器。图 13.14 给出了一个结构体类型 Student 的定义。在结构体 Student 中定义了 3 个构造器。其中,在第三个构造器的实现中,通过 self.init 调用类型本身已经定义的第二个构造器,实现对属性 name 和 age 的初始化。而在第三个构造器中,只需对多出的属性 id 赋值。

```
struct Student {
    var name : String
    var age : Int
    var id : String
    var basicInfo : String {
        return "\(name) is \(age) years old, the id is \(id)"
    }
    init(){
        name = "no name"
        age = 16
        id = ""
    }
    init(name : String, age : Int){
        self.name = name
        self.age = age
        id = ""
    }
    init(name : String, age : Int, id : String){
        self.init(name : name, age : age)
        self.id = id
    }
}

let theStudent = Student(name: "Sam", age: 28, id: "BY0601115")
```

```
name "Sam"
age 28
id "BY0601115"
```

图 13.14 值类型的构造器代理

13.3.8 引用类型的构造器代理

类是引用类型。引用类型里的所有存储属性,包括从父类继承的属性,都必须在构造器中进行初始化。有两种类型的引用类型构造器,分别为指定构造器和便利构造器。

指定构造器是类中最主要的构造器,它负责初始化类中定义的所有属性,并调用父类的构造器实现父类中属性的初始化。类的定义中必须有一个指定构造器。

便利构造器是类中一个重要的辅助构造器。它可以调用同一个类中的指定构造器。

指定构造器的定义格式为:

```
init(parameters list) {
some statements
}
```

便利构造器的格式和指定构造器的格式类似,只是在 init 前面多了一个关键字 convenience,具体格式为:

```
convenience  init(parameters list) {
some statements
}
```

对于指定构造器和便利构造器的使用,Swift 中有 3 个规则:

(1) 指定构造器必须调用其直接父类的指定构造器。
(2) 便利构造器必须调用同一个类中的其他构造器。
(3) 便利构造器必须最终以调用一个指定构造器结束。

这 3 条规则可以理解为:指定构造器是纵向代理,即向父类代理。便利构造器则是横向代理,即在同一个类中的构造器代理。

在 Swift 中,子类是不会自动继承父类构造器的,需要使用关键字 override 重载父类构造器。

如图 13.15 所示,父类 Student 中有两个构造器:一个为指定构造器,不带参数;另一个为便利构造器,含有 3 个参数。在便利构造器的实现中,必须通过 self.init 调用其指定构造器。

```
class Student {
    var name : String
    var age : Int
    var id : String
    var basicInfo : String {
        return "\(name) is \(age) years old, the id is \(id)"
    }
    init(){
        name = "no name"
        age = 16
        id = ""
    }
    convenience init(name : String, age : Int, id : String){
        self.init()
        self.name = name
        self.age = age
        self.id = id
    }
}
```

图 13.15　默认构造器和便利构造器

图 13.16 定义了 Student 类的子类 Graduate。其中定义了两个构造器：第一个构造器重载了父类的构造器，所以加上了关键字 override，并在实现中调用了父类的指定构造器；第二个构造器是便利构造器，通过调用同一个类中的构造器初始化了部分属性。

```
class Graduate : Student {
    var supervisor : String
    var researchTopic : String
    override init() {
        supervisor = ""
        researchTopic = ""
        super.init()
    }
    convenience init(name : String, age : Int, id : String,supervisor : String,researchTopic : String) {
        self.init(name : name, age : age, id : id)
        self.supervisor = supervisor
        self.researchTopic = researchTopic
    }
}
```

图 13.16　重载父类构造器

图 13.17 显示了子类 Graduate 实例创建后的所有属性的初始化值。

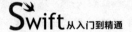

```
var theGraduate = Graduate(name: "Tom", age: 29, id: "BY0602115",
    supervisor: "Ian", researchTopic: "Petri Net")
▶{name "Tom", age 29, id "BY0602115"}
 supervisor "Ian"
 researchTopic "Petri Net"
```

图 13.17　实例初始化

练习题

1. 在 13.2 节练习题的基础上，分别为类 Vehicle 和类 Car 增加默认构造器，并去除属性定义部分的赋初始值的语句。

2. 继续在类 Vehicle 和类 Car 中增加带参数的自定义构造器，要求自定义构造器的参数涵盖所有类中的属性。

3. 在类 Car 中增加一个便利构造器，该构造器有一个参数，为 Car 类类型。定义一个 Car 类的实例 myCar，并初始化。然后，将其 owner 属性修改为 Jim。再定义一个常量 hisCar，调用 Car 类的便利构造器赋值给 hisCar，参数为 myCar。

13.4　类的析构

只有引用类型的类才有析构器。当一个类的实例使用结束，要被释放之前，析构器就会被调用。定义析构器要使用关键字 deinit，定义方式和定义构造器的方式类似。每个类最多只能有一个析构器。析构器不带任何参数，具体格式如下：

```
deinit {
    //release some resource
}
```

Swift 会自动释放不再需要的实例资源，而不需要手动清理。但是，如果实例使用了自己申请的资源，就需要在析构器里进行一些额外的清理工作了。例如，创建了一个自定义的类打开文件，并写入一些数据。这种情况就需要通过析构器在类实例被释放之前，手动关闭文件。

析构器不需要主动调用。在系统释放实例资源前，析构器会被自动调用。子类会继承父类的析构器。在子类析构器实现的最后，父类的析构器会被自动调用。如果子类没有定义自己的析构器,父类的析构器也会被自动调用。由于析构器被调用时,实例资源还没有被释放,所以析构器可以访问所有请求实例的属性。

如图 13.18 和图 13.19 所示,在上例的基础上,为父类 Student 和子类 Graduate 增加析构器。在析构器里打印该类析构器被调用的信息,以模拟释放实例占用的资源。

```
class Student {
    var name : String
    var age : Int
    var id : String
    var basicInfo : String {
        return "\(name) is \(age) years old, the id is \(id)"
    }
    init(){
        name = "no name"
        age = 16
        id = ""
    }
    convenience init(name : String, age : Int, id : String){
        self.init()
        self.name = name
        self.age = age
        self.id = id
    }
    deinit {
        print("call deinit of Class Student")
    }
}
```

图 13.18 父类 Student 中增加析构器

```
class Graduate : Student {
    var supervisor : String
    var researchTopic : String
    override init() {
        supervisor = ""
        researchTopic = ""
        super.init()
    }
    convenience init(name : String, age : Int, id : String,supervisor :
        String,researchTopic : String) {
        self.init(name : name, age : age, id : id)
        self.supervisor = supervisor
        self.researchTopic = researchTopic

    }
    deinit {
        print("call deinit of Class Graduate")
    }
}
```

图 13.19 子类 Graduate 中增加析构器

如图 13.20 所示,创建一个 Graduate 类的实例,该实例为可选型。然后,将 nil 赋值给实例 theGraduate。此时,系统首先调用基类 Student 的析构器,然后调用子类 Graduate 的析构器,最后调用自动释放实例资源的程序。

```
45  var theGraduate : Graduate? = Graduate(name : "Tom", age : 29, id :
        "BY0602115",supervisor: "Ian", researchTopic: "Petri net")
46  theGraduate = nil
```

```
call deinit of Class Graduate
call deinit of Class Student
```

图 13.20　析构器运行结果

本章知识点

1. 继承性相关的知识点,包括基类的定义和使用,子类和父类的定义和使用以及子类的多态性。

2. 重载相关的知识点,包括重载方法、重载属性、向重载属性中增加观察器、禁止重载。

3. 类的构造相关的知识点,包括构造器的定义和无参数默认构造器;带参数的自定义构造器,无外部参数名的自定义构造器,可选型属性的自动初始化,常量属性的初始化,值类型构造器代理,引用类型的构造器代理、指定构造器和便利构造器。

4. 析构器的定义和使用。

参考代码

【13.1 节　继承性】

第 1 题的参考代码-继承性如图 13.21 所示。
第 2 题的参考代码-继承性如图 13.22 所示。
第 3 题的参考代码-继承性如图 13.23 所示。

```
class Vehicle {
    var use: String = "No use"
    var speed: Int = 0
    var producer: String = "No producer"
    var usedYear: Int = 0
    var basicInfo: String {
        return "use:\(use),speed:\(speed),producer:\(producer)"
    }
    func description(){
        print("Vehicle's description is:")
        print("use:\(use),speed:\(speed),producer:\(producer)")
    }
}

let train = Vehicle()
train.use = "Common"
train.speed = 60
train.producer = "CRRC"
train.usedYear = 3
train.description()
```

图 13.21　第 1 题的参考代码-继承性

```
class Car: Vehicle {
    var owner: String = "No owner"
    var license: String = "No license"
    func propertyTransfer(newOwner: String){
        owner = newOwner
        print("The new owner is \(owner)")
    }
}

let myCar = Car()
myCar.owner = "Tommy"
myCar.license = "BJ0515G"
myCar.use = "Personal"
myCar.speed = 120
myCar.producer = "Tesla"
myCar.usedYear = 6
myCar.description()
myCar.propertyTransfer(newOwner: "Herry")
```

图 13.22　第 2 题的参考代码-继承性

```
func annualInspection(vehicle: Vehicle) -> Bool {
    if vehicle.usedYear < 5 {
        print("Needn't annual inspection")
        return false
    } else {
        print("Need to inspect")
        return true
    }
}

annualInspection(vehicle: train)
annualInspection(vehicle: myCar)
```

图 13.23　第 3 题的参考代码-继承性

【13.2 节　重载】

第 1 题的参考代码-重载如图 13.24 所示。

```swift
class Car: Vehicle {
    var owner: String = "No owner"
    var license: String = "No license"
    func propertyTransfer(newOwner: String){
        owner = newOwner
        print("The new owner is \(owner)")
    }
    override func description() {
        print("Car's description is:")

        print("use:\(use),speed:\(speed),producer:\
        (producer),usedYear:\(usedYear),owner:\
        (owner),license:\(license)")
    }
}

let myCar = Car()
myCar.owner = "Tommy"
myCar.license = "BJ0515G"
myCar.use = "Personal"
myCar.speed = 120
myCar.producer = "Tesla"
myCar.usedYear = 6
myCar.description()
```

图 13.24　第 1 题的参考代码-重载

第 2 题的参考代码-重载如图 13.25 所示。

```swift
class Car: Vehicle {
    var owner: String = "No owner"
    var license: String = "No license"
    func propertyTransfer(newOwner: String){
        owner = newOwner
        print("The new owner is \(owner)")
    }
    override func description() {
        print("Car's description is:")

        print("use:\(use),speed:\(speed),producer:\
        (producer),usedYear:\(usedYear),owner:\
        (owner),license:\(license)")
    }
```

图 13.25　第 2 题的参考代码-重载

```
        override var basicInfo: String {
            return super.basicInfo + "
                owner:\(owner),license:\(license)"
        }
    }

    let myCar = Car()
    myCar.owner = "Tommy"
    myCar.license = "BJ0515G"
    myCar.use = "Personal"
    myCar.speed = 120
    myCar.producer = "Tesla"
    myCar.usedYear = 6
    myCar.description()
    myCar.basicInfo
```

图 13.25 （续）

第 3 题的参考代码-重载如图 13.26 所示。

```
class Car: Vehicle {
    var owner: String = "No owner"
    var license: String = "No license"
    func propertyTransfer(newOwner: String){
        owner = newOwner
        print("The new owner is \(owner)")
    }
    override func description() {
        print("Car's description is:")

            print("use:\(use),speed:\(speed),producer:\
            (producer),usedYear:\(usedYear),owner:\
            (owner),license:\(license)")
    }
    override var basicInfo: String {
        return super.basicInfo + "
            owner:\(owner),license:\(license)"
    }
    override var usedYear: Int {
        didSet {
            print("Used Year was changed from \(oldValue)
                to \(usedYear)")
        }
        willSet {
            print("Used Year will be changed to
                \(newValue)")
        }
    }
}
```

图 13.26　第 3 题的参考代码-重载

```
let myCar = Car()
myCar.owner = "Tommy"
myCar.license = "BJ0515G"
myCar.use = "Personal"
myCar.speed = 120
myCar.producer = "Tesla"
myCar.usedYear = 6
myCar.description()
myCar.basicInfo
myCar.usedYear = 7
```

图 13.26 （续）

【13.3 节 类的构造】

第 1 题的参考代码-类的构造如图 13.27 所示。

```
class Vehicle {
    var use: String
    var speed: Int
    var producer: String
    var usedYear: Int
    var basicInfo: String {
        return "use:\(use),speed:\(speed),producer:\(producer)"
    }
    func description(){
        print("Vehicle's description is:")
        print(basicInfo)
    }
    init() {
        use = "No use"
        speed = 0
        producer = "No producer"
        usedYear = 0
    }
}
class Car: Vehicle {
    var owner: String
    var license: String
    func propertyTransfer(newOwner: String){
        owner = newOwner
        print("The new owner is \(owner)")
    }
    override func description() {
        print("Car's description is:")
```

图 13.27 第 1 题的参考代码-类的构造

```
            print("use:\(use),speed:\(speed),producer:\
                (producer),usedYear:\(usedYear),owner:\(owner),license:\
                (license)")
    }
    override var basicInfo: String {
        return super.basicInfo + " owner:\(owner),license:\(license)"
    }
    override var usedYear: Int {
        didSet {
            print("Used Year was changed from \(oldValue) to
                \(usedYear)")
        }
        willSet {
            print("Used Year will be changed to \(newValue)")
        }
    }
    override init() {
        owner = "No owner"
        license = "No license"
        super.init()
    }
}
```

图 13.27 （续）

第 2 题的参考代码-类的构造如图 13.28 所示。

```
        init(use: String, speed: Int, producer: String, usedYear: Int) {
            self.use = use
            self.speed = speed
            self.producer = producer
            self.usedYear = usedYear
        }
    }

let train = Vehicle(use: "Common", speed: 60, producer: "CRRC",
    usedYear: 3)

        init(use: String, speed: Int, producer: String, usedYear: Int,
            owner: String, license: String) {
            self.owner = owner
            self.license = license
            super.init(use: use, speed: speed, producer: producer,
                usedYear: usedYear)
        }
    }

let myCar = Car(use: "Personal", speed: 120, producer: "Tesla",
    usedYear: 6, owner: "Tommy", license: "BJ0515G")
```

图 13.28　第 2 题的参考代码-类的构造

第 3 题的参考代码-类的构造如图 13.29 所示。

```swift
convenience init(car: Car) {
    self.init(use: car.use, speed: car.speed, producer:
        car.producer, usedYear: car.usedYear, owner: car.owner,
        license: car.license)
}

let myCar = Car(use: "Personal", speed: 120, producer: "Tesla",
    usedYear: 6, owner: "Tommy", license: "BJ0515G")
myCar.owner = "Jim"
let hisCar = Car(car: myCar)
```

图 13.29　第 3 题的参考代码-类的构造

第 14 章 协议

协议可以理解为一个接口。协议中定义了为实现特定功能而需要的属性和方法，并不提供具体方法的实现。类、结构体、枚举型都可以通过实现协议中定义的方法遵守协议。另外，协议本身也可以通过扩展增加新的功能，从而得到一个新的协议。

14.1 协议的使用

14.1.1 协议的定义

协议的定义格式如下：

```
protocol theProtocol {
detailed definition
}
```

协议的定义和类的定义类似，只是使用了不同的关键字 protocol。要声明一个类遵守某个协议，需要在类声明中的类名后面加上协议的名称，中间用冒号":"隔开。当一个类遵守多个协议时，各协议之间用逗号分隔。如下所示，类 theClass 遵守协议 Protocol1 和 Protocol2。

```
class theClass : Protocol1, Protocol2 {
detailed definition of class
}
```

14.1.2 类遵守协议的声明

如果一个类既有父类,又要遵守协议,则需要将父类的名称放在协议名称的前面,并以逗号分隔。如下例所示,theClass 的父类 theFatherClass 必须放在协议的前面。

```
class theClass: theFatherClass, Protocol1, Protocol2 {
    detailed definition of class
}
```

14.1.3 协议中的属性

在协议中定义实例属性或者类属性时,需要提供属性的名称和类型,同时还必须指明是只读的,还是读写的。协议中通常用变量声明实例属性,并在类型声明后加上 {get set} 表示属性是只读的,还是可读写的。例如,{get} 是只读的,{get set} 是可读写的。在协议中定义类属性时,使用 var 关键字标示,具体格式如下:

```
protocol theProtocol {
    var readAndWriteAttribute : Int { get set }
    var onlyReadAttribute : Int { get }
}
```

当一个类遵守某个协议时,可以用 class 或者 static 关键字声明这个类属性,具体格式为

```
protocol theProtocol {
    static var onlyReadAttribute: Int { get }
}
```

如图 14.1 所示,定义了一个协议 Person,该协议规定了两个属性,分别为字符串型读写属性 name 和整型只读属性 age。另外,定义了 Student 类,该类遵守 Person 协议。因此,类 Student 的定义中必须包含协议 Person 中的属性 name 和 age。

```
protocol Person {
    var name: String { get set }
    var age: Int { get }
}
class Student: Person {
    var name : String
    var age : Int
    init(){
        name = ""
        age = 0
    }
}
```

图 14.1 协议的定义

14.1.4 协议中的关联类型

如果在定义协议属性的时候无法确定属性被实现时的具体类型或者属性在不同的实现时可能有不同的类型,这就需要使用协议中的关联类型了。

协议中关联类型定义的具体格式如下:

```
protocol theProtocol {
    associatedtype AttributeType
    var attribute: AttributeType { get }
}
```

其中,associatedtype 为关键字,AttributeType 为属性类型的名称,attribute 为属性。

通过协议中的关联类型,可以在遵守该协议的类中确定具体的类型。如图 14.2 所示,在协议 Person 中增加了两行语句:一行为关联类型 UnknownType 声明;另一行为属性 weight 的定义,通过 UnknownType 声明其类型。类 Student 遵守协议 Person,声明了协议中的所有存储属性。在声明属性 weight 时,根据实际的精度要求将其类型设置为 Double,这里也可以设置任何其他的类型。

```
protocol Person {
    associatedtype UnknownType
    var name: String { get set}
    var age: Int { get }
    var weight: UnknownType { get }
}

class Student: Person {
    var name : String
    var age : Int
    var weight: Double
    init(){
        name = ""
        age = 0
        weight = 0.0
    }
}
```

图 14.2 协议中的关联类型

14.1.5 协议中的方法

在协议中定义实例方法或类方法时,不需要花括号和方法体,只提供方法名称及参数列表、返回值即可,具体格式如下:

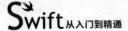

```
protocol theProtocol {
func methodName(parameter list)->ReturnType
}
```

如果定义的是类方法,则要使用关键字 static 标明。遵守该协议的类在定义实现方法时,也要使用关键字 class 或 static,具体格式如下:

```
protocol theProtocol {
static func methodName(parameter list)->ReturnType
}
```

如图 14.3 所示,在上例的基础上,为协议 Person 增加一个方法的声明(不含方法体)。在遵守该协议的类 Student 中需要实现这个方法,本例中的实现是打印学生描述信息。

```
protocol Person {
    associatedtype UnknownType
    var name : String { get set}
    var age : Int { get }
    var weight: UnknownType { get }
    func description()
}

class Student: Person {
    var name : String
    var age : Int
    var weight: Double
    func description() {
        print("\(name) is \(age)")
    }
    init(){
        name = ""
        age = 0
        weight = 0.0
    }
}
```

图 14.3 协议的实例

14.1.6 协议中的构造器

在协议中还可以声明构造器,遵守协议的类必须实现该构造器。协议中的构造器不需要写花括号和构造器实现。具体格式如下:

```
protocol theProtocol {
init (parameterlist)
}
```

在遵守该协议的类中必须实现协议中声明的构造器,同时还必须写上关键字 required,具体格式如下:

```
class theClass : theProtocol {
    required init (parameterlist) {
        implementation
    }
}
```

如图 14.4 所示,在上例的基础上,协议 Person 中增加了构造器的声明。在遵守该协议的类 Class 中增加了构造器 init 的实现,并在构造器名字前增加了关键字 required。

```
protocol Person {
    associatedtype UnknownType
    var name : String { get set }
    var age : Int { get }
    var weight: UnknownType { get }
    func description()
    init(name : String, age : Int, weight: UnknownType)
}

class Student: Person {
    var name : String
    var age : Int
    var weight: Double
    func description() {
        print("\(name) is \(age)")
    }
    required init(name: String, age: Int, weight: Double) {
        self.name = name
        self.age = age
        self.weight = weight
    }
    init(){
        name = ""
        age = 0
        weight = 0
    }
}
```

图 14.4 带构造器的协议

14.1.7 扩展类实现协议

除了直接在类的声明中标明遵守某个协议,也可以通过扩展的方式声明一个类遵守协议,其效果和在类定义中声明一样。如图 14.5 所示。在协议 Person 中声明了一个 description() 方法。在类 Student 的声明中没有涉及协议。在不修改类 Student 声明的情况下,要使类 Student 遵守协议 Person,可以通过扩展的方式使类 Student 遵守协议 Person。通过扩展方式遵守协议,就要在扩展中实现协议中的方法 description()。扩展了类 Student 后,创建一个 Student 实例 theStudent,并通过实例调用协议中的方法 description(),运行结果正确。

```
protocol Person {
    func description()
}
class Student{
    var name : String
    var age : Int
    var weight: Double
    init(){
        name = ""
        age = 0
        weight = 0.0
    }
    init(name: String, age: Int, weight: Double) {
        self.name = name
        self.age = age
        self.weight = weight
    }
}
extension Student : Person {
    func description() {
        print("\(name) is \(age)")
    }
}
let theStudent = Student(name: "Tom",age: 23, weight: 69.2)
theStudent.description()
```

```
name "Tom"
age 23
weight 69.2
```

图 14.5　扩展类实现协议

本例中,将协议 Person 中的存储属性都移动到类 Student 的定义中了。因为 Swift 语法规定通过扩展类实现协议时,不能声明协议中的存储属性。

14.2 协议的继承性

一个协议可以继承一个或者多个其他协议,并在其基础上增加新的内容。协议继承的具体格式如下:

```
protocol newProtocol : Protocol1, Protocol2 {
new features
}
```

如图 14.6 所示,定义了 3 个协议 Person、Student 和 Graduate。其中,协议 Person 定义了两个读写属性 name、age 和一个方法 personDescription();协议 Student 定义了一个读写属性 school 和一个方法 studentDescription();协议 Graduate 继承了协议 Person 和 Student,另外还定义了读写属性 supervisor 和方法 graduateDescription()。

```
protocol Person {
    associatedtype UnknownType
    var name : String {get set}
    var age : Int {get set}
    var weight: UnknownType { get }
    func personDescription()
}

protocol Student {
    var school : String {get set}
    func studentDescription()
}

protocol Graduate : Person, Student {
    var supervisor : String {get set}
    func graduateDescription()
}
```

图 14.6 协议的继承性实例

如图 14.7 所示,定义了类 ComputerAssociationMember,该类遵守协议 Graduate。而协议 Graduate 又继承了协议 Person 和 Student,所以该类要实现这 3 个协议的内容,包括属性 name、age、weight、school、supervisor 和方法 personDescripion()、studentDescription()、graduateDescription()。创建类 ComputerAssociationMember 的实例 theMember,并通过实例调用协议中定义的方法。

```
class ComputerAssociationMember : Graduate {
    var name : String = ""
    var age : Int
    var weight: Double
    var school : String
    var supervisor : String
    func personDescription() {
        print("It's person description")
    }
    func studentDescription() {
        print("It's student description")
    }
    func graduateDescription() {
        print("It's graduate description")
    }
    init(name : String, age : Int, weight: Double, school : String,
        supervisor :String){
        self.name = name
        self.age = age
        self.weight = weight
        self.school = school
        self.supervisor = supervisor
    }
}

let theMember = ComputerAssociationMember(name: "Tom", age: 23,
    weight: 69.2, school: "BUAA", supervisor: "Ian")
theMember.personDescription()
theMember.studentDescription()
theMember.graduateDescription()
```

图 14.7 遵守继承协议的实例

练习题

1. 定义协议 Ownership，用来表示产权。该协议包括属性：物主 owner、产权登记日 registerDate、序列号 serialNo。另外，该协议还声明了一个方法：更新物主 updateOwnership()，含有新物主和登记日两个参数。

2. 定义协议 Vehicle，用来表示交通工具。该协议包括属性：号牌 license、生产商 producer、时速 speed、累计小时数 hour、载重量 weight（类型在实现时确定）、里程 distance（读写属性）。另外，该协议还声明了一个方法：交通工具行驶 movingVehicle()，含有一个参数，即新增行驶小时数。

3. 定义汽车类 Car，并声明遵守协议 Ownership、Vehicle。在类定义中，声明协议中的属性（里程 distance 按照计算属性实现），并实现相关方法。另外，定义类的默认构造器和带参数的自定义构造器。

4. 定义类 Car 的实例 myCar,并调用类构造器对实例进行初始化。调用实例方法模拟汽车新增行驶 800 小时,并更换车主。

5. 用扩展类的方法遵守协议 Ownership 和 Vehicle。这里先将协议 Ownership 和 Vehicle 中的存储属性都删掉,同时在类 Car 中增加这些存储属性。现在两个协议中都只有一个方法了。修改好协议和类之后,通过扩展类 Car,使其遵守协议 Ownership 和 Vehicle。最后,按照第 4 题的要求创建实例并调用相应的方法。

6. 定义协议 ConsumerGoods,用来表示消费品。该协议包括属性:价格 price、类型 type、生产日期 produceDate。另外,该协议还声明了一个方法:打印消费品信息 consumerGoodsInfo()。

7. 定义协议 ElectricAppliance,用来表示电气设备。该协议包括属性:功率 power、能耗级别 energyConsumptionLevel、用途 use。另外,该协议还声明了一个方法:打印电气设备信息 electricApplianceInfo()。

8. 定义协议 DomesticAppliance,用来表示家用电器。该协议遵守协议 ConsumerGoods 和 ElectricAppliance,并包括属性:商品可打折扣 discount、安装位置 location。另外,该协议还声明了一个方法:打印家用电器信息 domesticApplianceInfo()。

9. 定义空调类 AirConditioner。该类遵守协议 DomesticAppliance。在类定义中声明协议中的属性,并实现相关方法。另外,定义一个类构造器。定义空调类的一个实例,对其进行初始化,并调用实例的所有方法。

本章知识点

1. 协议的定义,类遵守协议的声明,协议中的属性、方法和构造器的定义和使用,以及通过扩展类实现协议的方法。

2. 协议的继承性的定义和应用。

参考代码

第 1 题的参考代码如图 14.8 所示。

```swift
protocol Ownership {
    var owner: String{ get }
    var registerDate: Date{ get }
    var serialNo: String{ get }
    func updateOwnership(newOwner: String, registerDate: Date)
}
```

图 14.8　第 1 题的参考代码

第 2 题的参考代码如图 14.9 所示。

```swift
protocol Vehicle {
    associatedtype WeightType
    var license: String{ get }
    var producer: String{ get }
    var speed: Int{ get }
    var hour: Int{ get }
    var distance: Int{ get set }
    var weight: WeightType { get }
    func movingVehicle(addedhour: Int)
}
```

图 14.9　第 2 题的参考代码

第 3 题的参考代码如图 14.10 所示。

```swift
class Car : Ownership, Vehicle {
    var owner: String
    var registerDate: Date
    var serialNo: String
    var license: String
    var producer: String
    var speed: Int
    var hour: Int
    var weight: Double
    var distance: Int {
        get {
            return speed * hour
        }
        set {
            hour = distance / speed
        }
    }
    func updateOwnership(newOwner: String, registerDate: Date) {
        owner = newOwner
        self.registerDate = registerDate
    }
    func movingVehicle(addedhour: Int) {
        hour += addedhour
```

图 14.10　第 3 题的参考代码

```
    }
    init() {
        owner = ""
        registerDate = Date()
        serialNo = ""
        license = ""
        producer = ""
        speed = 0
        hour = 0
        weight = 0.0
    }
    init(owner: String, registerDate: Date, serialNo: String, li
        String, producer: String, speed: Int, hour: Int, weight:
        self.owner = owner
        self.registerDate = registerDate
        self.serialNo = serialNo
        self.license = license
        self.producer = producer
        self.speed = speed
        self.hour = hour
        self.weight = weight
    }
}
```

图 14.10　第 3 题的参考代码(续)

第 4 题的参考代码如图 14.11 所示。

```
let myCar = Car(owner: "Tom", registerDate: Date(), serialNo:
    "102008930", license: "BJ0515G", producer: "BMW", speed: 150, hour:
    0, weight: 3.5)
myCar.movingVehicle(addedhour: 800)
myCar.updateOwnership(newOwner: "Sam", registerDate:
    Date(timeIntervalSinceNow: 10000))
```

图 14.11　第 4 题的参考代码

第 5 题的参考代码如图 14.12 所示。

```
protocol Ownership {
    func updateOwnership(newOwner: String, registerDate: Date)
}

protocol Vehicle {
    func movingVehicle(hour: Int)
}
class Car {
    var owner: String
    var registerDate: Date
    var serialNo: String
    var license: String
    var producer: String
    var speed: Int
    var hour: Int
```

图 14.12　第 5 题的参考代码

```swift
        var weight: Double
        var distance: Int {
            get {
                return speed * hour
            }
            set {
                hour = distance / speed
            }
        }
        init(owner: String, registerDate: Date, serialNo: String,
            license: String, producer: String, speed: Int, hour: Int,
            weight: Double) {
            self.owner = owner
            self.registerDate = registerDate
            self.serialNo = serialNo
            self.license = license
            self.producer = producer
            self.speed = speed
            self.hour = hour
            self.weight = weight
        }
}

extension Car: Ownership, Vehicle {
    func updateOwnership(newOwner: String, registerDate: Date) {
        owner = newOwner
        self.registerDate = registerDate
    }
    func movingVehicle(hour: Int) {
        self.hour += hour
    }
}

let myCar = Car(owner: "Tom", registerDate: Date(), serialNo:
    "102008930", license: "BJ0515G", producer: "BMW", speed: 150,
    hour: 0, weight: 3.5)
myCar.movingVehicle(hour: 800)
myCar.updateOwnership(newOwner: "Sam", registerDate:
    Date(timeIntervalSinceNow: 10000))
myCar.distance
```

图 14.12 （续）

第 6 题的参考代码如图 14.13 所示。

```swift
protocol ConsumerGoods{
    var price: Double { get }
    var type: String { get }
    var produceDate: Date { get }
    func consumerGoodsInfo()
}
```

图 14.13　第 6 题的参考代码

第 7 题的参考代码如图 14.14 所示。

```swift
protocol ElectricAppliance{
    var power: Int { get }
    var energyConsumptionLevel: Int { get }
    var use: String { get }
    func electricApplianceInfo()
}
```

图 14.14　第 7 题的参考代码

第 8 题的参考代码如图 14.15 所示。

```swift
protocol DomesticAppliance: ConsumerGoods, ElectricAppliance {
    var discount: Double { get }
    var location: String { get }
    func domesticApplianceInfo()
}
```

图 14.15　第 8 题的参考代码

第 9 题的参考代码如图 14.16 所示。

```swift
class AirConditioner: DomesticAppliance {
    var price: Double
    var type: String
    var produceDate: Date
    var power: Int
    var energyConsumptionLevel: Int
    var use: String
    var discount: Double
    var location: String
    
    func consumerGoodsInfo() {
        print("Price:\(price), Type:\(type), Produce
            Date:\(produceDate)")
    }
    
    func electricApplianceInfo() {
        print("Power:\(power), energy consumption
            level:\(energyConsumptionLevel), use:\(use)")
    }
    
    func domesticApplianceInfo() {
        print("Discount:\(discount), location:\(location)")
    }
    
    init(price: Double, type: String, produceDate: Date, power:
        Int, energyConsumptionLevel: Int, use: String, discount:
        Double, location: String) {
```

图 14.16　第 9 题的参考代码

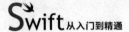

```
            self.price = price
            self.type = type
            self.produceDate = produceDate
            self.power = power
            self.energyConsumptionLevel = energyConsumptionLevel
            self.use = use
            self.discount = discount
            self.location = location
        }
    }
57  let myAirConditioner = AirConditioner(price: 3800, type:
        "small", produceDate: Date(), power: 2000,
        energyConsumptionLevel: 2, use: "Civil use", discount: 0.6,
        location: "Outdoors")
58  myAirConditioner.consumerGoodsInfo()
59  myAirConditioner.electricApplianceInfo()
60  myAirConditioner.domesticApplianceInfo()
```

```
Price:3800.0, Type:small, Produce Date:2019-07-02 11:20:41 +0000
Power:2000, energy consumption level:2, use:Civil use
Discount:0.6, location:Outdoors
```

图 14.16 （续）

第 15 章 泛 型

泛型是 Swift 的一个重要特征,它是一种适用于任何类型的可重用函数或类型,可以避免编写重复的代码。Swift 标准库就是在泛型的基础上构建的。例如,最常使用的数组就是一种典型的泛型。数组既可以是整型的,也可以是字符串型的。

15.1 泛型函数

泛型函数是指函数中的参数可以是任何类型的。如图 15.1 所示,函数 swap2Int 的功能是交换两个整数的值。其中,两个参数均为 inout 整型。定义变量 a 和 b,并分别赋值为 3 和 5。然后通过函数交换 a 和 b 的值,并打印交换结果。

```
func swap2Int( a : inout Int, b : inout Int) {
    let temp = a
    a = b
    b = temp
}

var a = 3
var b = 5
swap(&a, &b)
print("a is \(a), b is \(b)")
```

图 15.1 swap2Int()函数的定义及实例

如图 15.2 所示,函数 swap2String 的功能是交换两个字符串的值。其中,两个参数均为 inout 字符串型。定义变量 c 和 d,并分别赋值为"Hello"和"world"。然后通过函数交换 c 和 d 的值,并打印交换结果。

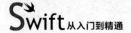

```
func swap2String( a : inout String, b : inout String) {
    let temp = a
    a = b
    b = temp
}
var c = "Hello"
var d = "world"
print("\(c) \(d)")
swap(&c, &d)
print("\(c) \(d)")
```

图 15.2　swap2String 函数及其定义

从功能上看，这两个函数都是交换两个变量的值，唯一的不同在于参数的类型。这种情况就可以通过泛型函数编写一个用于交换两个变量值的通用函数，变量的类型可以为任意类型。

如图 15.3 所示，在函数名后面加上"＜T＞"，表示类型名占位，用来替代具体的类型名。在函数的参数列表中，参数 a 和 b 的类型均为 T，说明 a 和 b 是同一种类型，但具体是什么类型还不知道。函数体的内容和前面两个函数一样，并没有任何改变。调用函数 swap2Element()时，当传入的参数为整数 a 和 b，则 T 表示整型；当传入的参数为字符串 c 和 d，则 T 表示字符串型；当参数为其他某个类型，那么 T 就是那个类型。通过泛型函数，可以在函数定义时不确定具体的参数类型，直到函数调用时才确定。这种方法大大增加了函数的通用性，避免了重复大量功能类似的代码。

```
print("a is \(a), b is \(b)")
print("c is \(c), d is \(d)")

func swap2Element<T>( a : inout T, b : inout T){
    let temp =  a
    a =  b
    b = temp
}

swap2Element(a: &a, b: &b)
swap2Element(a: &c, b: &d)
print("a is \(a), b is \(b)")
print("c is \(c), d is \(d)")
```

图 15.3　交换两个元素的泛型函数

占位类型一般使用 T 表示，也可使用其他任意名称。占位类型可以理解为一种类型的参数。类型参数可用来定义函数的参数类型和返回类型，也可以在函数体中使用。占位类型在函数被调用时，才会确定为一个具体的类型。

15.2 泛型类

泛型不仅可用于函数中,也可用于类和结构体中,其用法和泛型函数类似,表示在定义阶段无法确定的一种类型,这个类型在类实例化的时候才能确定。如图 15.4 所示,定义了两个交通工具的类,分别为汽车类 Car 和自行车类 Bike。这两个类是完全不一样的类,有不同的属性和方法。在此基础上又定义了一个驾驶员的类 Driver。这个类有一点特别,在类名后面有"＜Vehicle＞",表示该类为泛型类,即类中用到了泛型类别 Vehicle。这里,Vehicle 只是一个类型占位符,和前例中的 T 一样,但本例中使用 Vehicle 有助于理解。在 Driver 类中,变量 vehicle 表示驾驶员使用的交通工具。在定义类 Driver 的时候,无法确定具体的交通工具,直到类 Driver 实例化的时候才能确定。

```
class Car {
    var brand: String
    init(brand: String) {
        self.brand = brand
    }
}

class Bike {
    var speed: Int
    init(speed: Int) {
        self.speed = speed
    }
}

class Driver<Vehicle> {
    var name: String
    var vehicle: Vehicle
    init(name: String, vehicle: Vehicle) {
        self.name = name
        self.vehicle = vehicle
    }
}
```

图 15.4 泛型类

如图 15.5 所示,定义了两个交通工具的常量:porscheCar 为汽车类实例、giantBike 为自行车类实例。另外,还定义了两个驾驶员 driverTom 和 driverSam,在进行实例化的时候,分别指定汽车和自行车作为 vehicle 参数值,这时才确定了类

Driver 中泛型 Vehicle 具体的类型。

```
let porscheCar = Car(brand: "Porsche")
let driverTom = Driver(name: "Tom", vehicle: porscheCar)

let giantBike = Bike(speed: 50)
let driverSam = Driver(name: "Sam", vehicle: giantBike)
```

图 15.5　类实例化时确定泛型

泛型还可用于枚举型和结构体中，这里不再赘述。在 Swift 的库函数中大量使用了泛型，比较典型的实例有数组、字典和可选型。

练习题

1. 定义一个函数 swap2Element()，有两个参数 a 和 b，都不能确定类型。函数返回值为元组(b,a)。调用该函数，其中参数 a 为整数 33，参数 b 为字符串"Jay"，打印函数返回的元组。

2. 定义一个类 Person，含有 3 个属性：名字 name（字符型）、体重 weight（泛型）、身高 height（泛型）。类中有两个方法：一个为构造器；另一个为 description，负责打印 Person 类的属性信息。另外，定义 3 个实例，其中实例 jennie：名字为 Jennie、体重为 45、身高为 160；

实例 jack：名字为 Jack、体重为 69.2、身高为 173；

实例 sam：名字为 Sam、体重为 88.8、身高为 Unknown。

通过调用实例方法 description()，分别打印 3 个实例的相关信息。

本章知识点

1. 泛型的定义以及泛型函数的用法。
2. 泛型类的用法。

参考代码

第 1 题的参考代码如图 15.6 所示。

第 15 章 泛型

```swift
func swap2Element<T, U>(a: T, b: U) -> (U, T) {
    return (b, a)
}

let theTuple = swap2Element(a: 33, b: "Jay")
print(theTuple)
```

图 15.6　第 1 题的参考代码

第 2 题的参考代码如图 15.7 所示。

```swift
class Person<T, U> {
    var name: String
    var weight: T
    var height: U
    init(name: String, weight: T, height: U) {
        self.name = name
        self.weight = weight
        self.height = height
    }
    func description(){
        print("\(name)'s weight is \(weight) and height is \(height)")
    }
}

let jennie = Person(name: "Jennie", weight: 45, height: 160)
let jack = Person(name: "Jack", weight: 69.2, height: 173)
let sam = Person(name: "Sam", weight: 88.8, height: "Unknow")
jennie.description()
jack.description()
sam.description()
```

```
Jennie's weight is 45 and height is 160
Jack's weight is 69.2 and height is 173
Sam's weight is 88.8 and height is Unknow
```

图 15.7　第 2 题的参考代码

第 3 篇

深 入 篇

　　深入篇由7个章节构成,分别为异常、可选链、访问控制、类型操作符、扩展、内存管理以及高级运算符,这些主题内容上比较分散,都是前面章节在内容上的进一步深入。在学好前两篇的基础上,继续学习深入篇有助于全面系统地掌握Swift语言的知识体系,特别是更好地理解Swift语言的精髓。

第 16 章 异 常

异常是指在程序执行过程中出现的各种出错情况,如用户的错误输入、网络连接失败、外部文件访问异常等。异常处理就是当程序过程中发生了异常时,捕获异常并进行适当处理的过程。

16.1 定义异常

要处理异常,首先要知道什么是异常。在 Swift 中,用遵守协议 ErrorType 的类型表示异常。ErrorType 是一个空协议,专用于异常处理。

通常用枚举型定义一组相同类型的各种异常情形。图 16.1 定义了 3 种创建学生实例时可能出现的异常情形:没有名字、年龄不合法、ID 无效。

```
enum CreatingStudentError : Error {
    case NoName
    case InvalidAge
    case InvalidId
}
```

图 16.1 异常定义

16.2 抛出异常

在程序执行过程中发生异常时,通过抛出一个异常启动异常处理程序。如图 16.2 所示,当学生实例的年龄异常时,可以通过以下语句抛出一个年龄不合法 InvalidAge 的异常。当一个异常被抛出后,将由特定的异常处理程序接管,并对异常进行处理,如向用户提示异常信息。

$$\text{throw CreatingStudentError.InvalidAge}$$

图 16.2　抛出异常

要将函数中的异常传递给调用该函数的代码，就要用关键字 throws 标识这个可能抛出异常的函数，即在函数声明的参数列表后面加上 throws 关键字。一个标识了 throws 关键字的函数称为 throwing 函数。如果这个函数有返回值，那么 throws 关键字应该写在返回值符号"->"之前。throwing 函数从函数内部抛出异常，并将其传递到调用它的地方。

具体格式如下：

```
func canThrowError() throws->returnType{
    statements
}
```

图 16.3 定义了一个批量创建类 Student 的实例的函数。在函数声明的部分标注了关键字 throws，说明这是一个 throwing 函数。该函数在执行过程中如果发生了异常，会将异常传递到调用函数的地方进行处理，而不是在函数内部进行处理。在该函

```
func createSomeStudents(counts : Int) throws -> [Student]{
    var studentsArray = [Student]()
    var i = counts
    while i > 0 {
        let theStudent = Student()
        guard theStudent.age > 15 else {
            throw CreatingStudentError.InvalidAge
        }
        guard theStudent.id.count == 8 else {
            throw CreatingStudentError.InvalidId
        }
        guard !theStudent.name.isEmpty else {
            throw CreatingStudentError.NoName
        }
        studentsArray.append(theStudent)
        i = i - 1
    }
    return studentsArray
}

let studentsArray = try createSomeStudents(counts: 5)
```

图 16.3　异常处理

数中,while 循环语句执行 counts 次循环,每次都创建一个默认的 Student 实例,然后对该实例中的 age、id、name 3 个属性的值进行检查。如果 age 为 15 以下或 id 字符串的长度不等于 8 或 name 字符串为空,都会导致抛出异常,并且中断程序继续向下执行。

16.3 处理异常

通过 do-catch 结构语句可以对各种异常情况进行逐一匹配和处理。可能抛出异常的语句要放到 do 语句中,然后通过 catch 语句匹配各种异常的类型,并进行相应的处理。具体格式为:

```
do {
try statement
} catch errorType1 {
statements
}catch errorType2 {
statements
}
```

如果 catch 语句中没有写出具体的 errorType,那么将匹配所有的异常类型。catch 语句不一定要处理 do 语句中抛出的所有可能的异常。如果 catch 语句没有匹配上异常类型,该异常会被传播到周围的作用域。

如图 16.4 所示,在上例的基础上,增加 do-catch 语句对函数中抛出的异常进行处理。

```
do {
    try createSomeStudents(counts: 1)
}catch CreatingStudentError.InvalidAge {
    print("Invalid age error!")
}catch CreatingStudentError.InvalidId {
    print("Invalid id error!")
}catch CreatingStudentError.NoName {
    print("No name error!")
}
```

图 16.4 异常处理实例

16.3.1 将异常定义为可选型

当对异常的具体情况并不关心时,可以将异常定义为可选型简化处理,即通过使用 try? 可以将异常转换为一个可选值处理。如果在 try? 语句中抛出了一个异常,那么这个表达式的值为 nil。图 16.5 定义了一个可能抛出异常的 throwing 函数。在常量 x 的赋值语句中,通过 try? 语句处理调用函数过程中可能发生的异常。如果在执行过程中抛出了异常,x 的值就为 nil。

```
func theThrowingFunction() throws -> Int {
    //some statements
    return 0
}
let x = try? theThrowingFunction()
```

图 16.5 将异常转换成一个可选值

16.3.2 断言异常不发生

已知某个 throwing 函数在运行时不会抛出异常,可以在表达式前用 try! 使异常传递失效。如果实际中抛出了异常,那么就会得到一个运行时异常。

练习题

定义了一个名为 NameError 的枚举型,其中只有一个值是 EmptyName,表示名字为空的异常。定义函数 canThrowError,其中含有一个名为 name 的字符串类型参数。该函数可以抛出异常。如果 name 为空,则抛出异常,否则打印出名字。在调用函数的时候,通过 do-catch 结构进行异常的捕获和处理。当函数抛出异常,之后进入 catch 块进行异常处理:打印异常信息"name is empty!"。两次调用函数:第一次调用函数时传入字符串"Tommy";第二次调用函数时传入空值。

本章知识点

1. 定义异常的方式。
2. 抛出异常的方式。
3. 处理异常的方式,包括:通过 do-catch 结构的代码处理异常、将异常定义为可选型处理、断言异常根本不会发生。

参考代码

练习题的参考代码如图 16.6 所示。

```
3   enum  NameError : Error {
4       case EmptyName
5   }
6
7   func canThrowError(name : String) throws {
8       if name.isEmpty {
9           throw NameError.EmptyName
10      }
11      print("There is no error")
12      print("The name is \(name)")
13  }
14
15  do {
16      try canThrowError(name: "Tommy")
17      try canThrowError(name: "")
18  } catch NameError.EmptyName {
19      print("There is an error")
20      print("name is empty!")
21  }
```

```
There is no error
The name is Tommy
There is an error
name is empty!
```

图 16.6 练习题的参考代码

第 17 章 可 选 链

可选链是调用可能为空的对象的属性、方法及下标的过程。如果对象不为空，则调用成功。如果对象为空，则返回 nil。多个连续的调用可以链接成一个调用链，其中任何一个节点的调用对象为 nil，都会导致整个链的调用失败。

17.1 可选链的定义

在可选型后面加上一个感叹号"!"可以强制拆包该可选型的值。强制拆包一般是在确定可选型中肯定有值时使用的，对一个值为 nil 的可选型强制拆包会导致系统崩溃。通过对可选型的强制拆包，使一个可选型重新变为一个非可选型。

一个可选链的调用是否成功，可以通过返回值是否为 nil 判断。可选链调用的实质是在原来调用结果类型上增加了可选型。例如，原来调用结果为 Int 型，则其可选链的调用结果就是 Int? 型。

如图 17.1 所示，类 Student 中有一个属性 majorChosen，表示已选的主修课，类型为可选型的 Major。类 Major 含有一个属性为 studentsNum 的整型变量，初始值为 65。创建 Student 类的实例 sam，其属性 majorChosen 为 nil。如果强制拆包实例 sam 的 majorChosen 属性的 studentsNum 属性，则会导致系统崩溃，因为不能强制拆包一个为 nil 的可选型。可选链提供了一种处理这个情况的方式，即在实例 sam 的可选型属性的后面加上问号"?"。这样，编译器就不会在该属性为 nil 的时候访问 studentsNum 属性了。在本例中，先通过可选链访问实例中的 studentsNum 属性，然后通过 let 语句与常量进行可选型绑定,并判断该常量的值:如果为 nil,就直接跳到 else 从句中执行;如果不为 nil,就继续获取 studentsNum 的值并赋值给常量,然后继续执行。

```
3   class Student {
4       var majorChosen : Major?
5   }
6
7   class Major {
8       var studentsNum = 65
9   }
10
11  let sam = Student()
12
13  //sam.majorChosen!.studentsNum
14
15  if let studentsNum = sam.majorChosen?.studentsNum {
16      print("students number is \(studentsNum).")
17  } else {
18      print("major chosen is nil!")
19  }
```

major chosen is nil!

图 17.1　可选链

17.2　可选链的使用

应用可选链可以实现向下访问多层的属性、方法及下标。下面通过由 4 个类构成的实例来说明。

图 17.2 定义了类 Student 和类 Major。

类 Student 只含有一个属性 majorChosen,为可选型,表示已选的主修课。

类 Major 表示主修的课程,它有 3 个属性:classrooms、classroomNum、location。其中,属性 classrooms 为数组类型,数组元素为类 Classroom 的实例。属性 classroomNum 为计算型属性,根据数组 classrooms 里的元素个数计算而得。属性 location 为类 Location 的可选型。另外,类中还定义了一个下标,用来根据教室号返回教室的实例或者为教室设置教室号。最后,类中定义了一个打印教室号的方法。

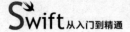

```
class Student {
    var majorChosen : Major?
}

class Major {
    var classrooms = [Classroom]()
    var classroomNum : Int {
        return classrooms.count
    }
    var location : Location?
    subscript(number : Int) -> Classroom {
        get {
            return classrooms[number]
        }
        set {
            classrooms[number] = newValue
        }
    }
    func ClassroomNumber() {
        print("Classroom number is \(classroomNum)")
    }
}
```

图 17.2　类 Student 和类 Major 的定义

图 17.3 定义了类 Classroom 和类 Location。

```
class Classroom {
    let name : String
    init(name : String) {
        self.name = name
    }
}

class Location {
    var buildingName : String?
    var buildingNumber : Int?
    var streetName : String?
    func buildingIdentifier() -> String? {
        if buildingName != nil {
            return buildingName
        }else if buildingNumber != nil {
            return "\(String(describing: buildingNumber))"
        }else {
            return nil
        }
    }
}
```

图 17.3　类 Classroom 和类 Location 的定义

类 Classroom 只有一个字符串属性 name，表示教室名。另外，还有一个构造方法，负责初始化教室名。

类 Location 表示位置，含有 3 个可选型字符串属性，分别为楼宇名 buildingName、楼宇号 buildingNumber、街道名 streetName。另外，还有一个返回楼宇标识的方法，返回类型为可选型字符串。

17.2.1 可选链访问实例属性

在此基础上，通过可选链访问实例属性，并通过可选型绑定判断结果是否为 nil。如图 17.4 所示，创建一个 Student 类的实例 sam。实例 sam 的 majorChosen 属性为可选型，因此访问 sam.majorChosen 中的属性 classroomNum 必须使用"?"，表示有可能为空。同样，访问 sam.majorChosen 中的属性 location 也必须使用"?"。

```
let sam = Student()
if let classroomNum = sam.majorChosen?.classroomNum {
    print("Classroom number is \(classroomNum).")
} else {
    print("Major chosen is nil")
}

let theLocation = Location()
theLocation.buildingNumber = 37
theLocation.streetName = "Xue Yuan Road"

sam.majorChosen?.location = theLocation
```
```
nil
```

图 17.4 可选链访问实例属性

17.2.2 可选链访问实例方法

通过可选链也可以访问实例方法，并判断访问是否成功。如图 17.5 所示，实例 sam 的属性 majorChosen 通过可选链调用实例方法 ClassroomNumber()，并通过 if 语句判断调用是否成功。

因为方法 ClassroomNumber() 没有返回值，所以返回结果为 Void，可选链 sam.majorChosen?.ClassroomNumber() 的类型就是"Void?"。本例中的调用实例方法虽然失败，但进行了恰当的处理。

```
if sam.majorChosen?.ClassroomNumber() != nil {
    print("Function ClassroomNumber() was called correctly!")
} else {
    print("Function call failed!")
}
```
```
Function call failed!
```

图 17.5　可选链访问实例方法

17.2.3　可选链访问实例下标

通过可选链还可以访问下标,并判断访问下标是否成功。这里需要注意的是,一般地,可选链的问号都是在可选表达式后面,但是访问下标时,可选链的问号要放在下标方括号的前面。如图 17.6 所示,可选链 sam.majorChosen?[0].name 通过下标访问数组元素 Classroom 类中的 name 属性,同样可以通过 if 语句判断访问该属性是否成功。

```
if let theNo1ClassroomName = sam.majorChosen?[0].name {
    print("The No.1 classroom is \(theNo1ClassroomName)")
} else {
    print("Subscript call failed!")
}
```
```
Subscript call failed!
```

图 17.6　应用可选链访问下标

练习题

1. 定义类 CarStore,表示车辆修理店。该类有两个属性:brand 表示修理店的品牌,为字符串型;address 表示修理店的地址,为字符串型。编写一个自定义的类构造器,对类中的属性进行初始化。

2. 定义类 Car,表示车辆。该类有两个属性:license 表示车牌照,为字符串可选型;repairShop 表示该车辆的维修店,为 CarStore 类型的数组。该类有一个方法 addNewRepairShop,该方法负责向维修店数组中添加一个新的维修店。另外,在该类中添加一个下标,用来设置或读取维修店数组中的某一个元素。

3. 定义类 Citizen，表示一个市民。该类有两个属性：name 表示名字，为字符串型；car 表示市民拥有的车辆，为 Car 类可选型。该类中有两个方法，其中方法 buyCar() 的参数为 Car 类型，负责设置 car 属性的值，方法 description() 在市民名下有车的时候打印出该市民的信息。

4. 定义一个 Citizen 实例 jerry。调用 jerry 的实例方法 description()，打印 jerry 名下的车辆信息。通过可选链访问并打印 jerry 名下车辆的车牌号。

5. 创建一个 Car 实例 newCar。创建一个 CarStore 实例 dasAutoStore。将 dasAutoStore 维修店添加到 newCar 的维修店中。jerry 通过方法 buyCar() 购买了一辆新车 newCar。再定义一个 CarStore 实例 audiStore。通过可选链调用实例方法，将 audiStore 添加到 jerry 的汽车维修店中，并打印相应信息。

6. 设置实例 jerry 的属性 car 的车牌照为 BJ0515G。然后，通过下标和 while 循环语句将 jerry 的车辆的所有维修店地址打印出来。

本章知识点

1. 可选链的定义。
2. 可选链的使用，包括：通过可选链访问实例属性、实例方法、实例下标。

参考代码

第 1 题的参考代码如图 17.7 所示。

```
class CarStore {
    var brand: String
    var address: String
    init(brand: String, address: String) {
        self.brand = brand
        self.address = address
    }
}
```

图 17.7　第 1 题的参考代码

第 2 题的参考代码如图 17.8 所示。
第 3 题的参考代码如图 17.9 所示。

```swift
class Car {
    var license: String?
    var repairShop: [CarStore]
    init() {
        repairShop = []
    }
    subscript(number: Int) -> CarStore {
        get {
            return repairShop[number]
        }
        set {
            repairShop[number] = newValue
        }
    }
    func addNewRepairShop(repairShop: CarStore) {
        self.repairShop.append(repairShop)
    }
}
```

图 17.8 第 2 题的参考代码

```swift
class Citizen {
    var name: String
    var car: Car?
    func description(){
        if car != nil {
            print("\(name) has a car. ")
        }
    }
    init(name: String) {
        self.name = name
    }
    func buyCar(car: Car) {
        self.car = car
    }
}
```

图 17.9 第 3 题的参考代码

第 4 题的参考代码如图 17.10 所示。

```swift
let jerry = Citizen(name: "Jerry")
jerry.description()
if let license = jerry.car?.license {
    print("jerry's car license is \(license)")
} else {
    print("jerry has no car or no license")
}
```

图 17.10 第 4 题的参考代码

第 5 题的参考代码如图 17.11 所示。

```
let newCar = Car()
let dasAutoStore = CarStore(brand: "Das Auto", address: "Beijing")
newCar.addNewRepairShop(repairShop: dasAutoStore)
jerry.buyCar(car: newCar)
let audiStore = CarStore(brand: "Audi", address: "Shanghai")
if jerry.car?.addNewRepairShop(repairShop: audiStore) != nil {
    print("add new repair store successfully!")
} else {
    print("fail to add new repair store!")
}
```

图 17.11　第 5 题的参考代码

第 6 题的参考代码如图 17.12 所示。

```
jerry.car?.license = "BJ0515G"
if let countOfCarStore = jerry.car?.repairShop.count {
    var i = 0
    while i < countOfCarStore {
        if let address = jerry.car?[i].address {
            print("No.1 car store lies in \(address)")
        }
        i += 1
    }
}
```

图 17.12　第 6 题的参考代码

第 18 章 访问控制

在 Swift 中，类、结构体和枚举类型都支持属性、方法和构造器。而属性、方法以及构造器可以看作这些类型的接口。随着代码量的增长，程序的复杂度越来越高，如何掌控系统中所有类型的接口是软件设计的关键。Swift 提供了访问控制的机制解决这个问题。

访问控制是用来限制源文件或模块中代码的访问级别，隐藏功能实现细节的一种机制。访问控制以声明的方式标注可以访问和使用的部分。通过访问控制，既可以设置类、结构体和枚举类型的访问级别，也可以设置类型的属性、函数、初始化方法、基本类型、下标等的访问级别。

18.1 访问级别

在介绍访问级别前，先介绍两个相关概念：模块和源文件。模块和源文件是 Swift 访问控制中的基本单位。

模块是以独立单元构建和发布的框架或应用。在 Swift 中可以通过关键字 import 引用一个模块。

源文件指的是 Swift 源代码的文件，它属于一个模块。

Swift 主要有 3 个访问级别，分别用关键字 public、internal 和 private 标注。

关键字 public 表示可以访问本模块中源文件中的任何实体，也可以通过引入该模块访问源文件中的所有实体。

关键字 internal 表示只能访问本模块中的任何 internal 实体，不能访问其他模块中的 internal 实体。internal 在使用时可以省略，因为 Swift 中的默认访问级别就是 internal。

关键字 private 表示只能在当前源文件中被使用的实体，也称为私有实体。private 可用来隐藏某些功能的实现细节。

在编写一个独立的应用程序时，所有功能都是为了实现该应用，不需要将代码提供给其他应用或者模块使用。在这种情况下，就不需要显式地设置访问级别，全部使用默认的 internal 访问级别即可，也就是不用进行访问级别标注。

但是，当开发一个框架时，需要把对外的接口定义为 public 级别，以便其他应用导入框架后可以正常调用框架中定义的实体。

在定义一个实体的访问级别时，需要在实体的定义前加上关键字 public、internal 和 private，并通过括号标注读写权限，具体格式如下：

private(set) var name: type

上面定义了类型为 type 的变量 name 的写权限为 private。由于 internal 为系统默认的访问级别，因此关键字 internal 在使用时可以省略。

18.2 实例

本节通过个人银行账号存取款的应用场景说明如何应用访问控制的机制解决实际问题。

如图 18.1 所示，定义一个银行账号的协议 BankAccount。所有银行账号相关的类都应该遵守这个协议，包括私人银行账号和公司银行账号等。为了简化，这个协议中只有 3 个属性和 3 个方法。属性包括：账号 accountID，可读，字符串型；户主 owner，可读，字符串型；账户余额 balance，可读，双精度浮点型。方法包括：存款 deposit（amount：Double）；取款 withdraw（amount：Double）；账户信息描述 description()。协议规定了各种银行账号都必须包含的属性和方法，没有对访问权限进行控制，只规定了属性的读写权限。

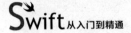

```
protocol BankAccount {
    var accountID: String { get }
    var owner: String { get }
    var balance: Double { get }
    func deposit(amount: Double)
    func withdraw(amount: Double)
    func description()
}
```

图 18.1　协议 BankAccount

图 18.2 定义了个人银行账户类 PrivateAccount，该类遵守协议 BankAccount。PrivateAccount 实现了协议中的所有方法。另外，该类还实现了自定义的构造器。该构造器通过库函数 UUID().uuidString 产生一个随机且唯一的字符串，作为账户 ID。

```
class PrivateAccount: BankAccount {
    var accountID: String
    var owner: String
    var balance: Double
    func deposit(amount: Double) {
        balance += amount
        print("deposit: \(amount), balance: \(balance)")
    }
    func withdraw(amount: Double) {
        if amount <= balance {
            balance -= amount
            print("withdraw: \(amount), balance: \(balance)")
        } else {
            print("balance = \(balance) is not enough for withdrawing \(amount) ")
        }
    }
    func description() {
        print("\(owner)'s Private Account: \(accountID), balance is \(balance)")
    }
    init(owner: String) {
        accountID = UUID().uuidString
        self.owner = owner
        balance = 0.0
    }
}
```

图 18.2　PrivateAccount 类

定义完私人银行账号类就可以进行存取款操作了。如图 18.3 所示，首先为用户 Tom 创建一个私人银行账号 PrivateAccount 实例，并调用构造器进行初始化。最后，向该账号存款 10 000 元，取款 8000 元。此时，打印账户信息显示余额为 2000 元。最后，直接向账户余额赋值 10 000 元，此时账户信息显示余额为 10 000 元。这个动作是

第 18 章 访问控制

非常危险的,因为它没有通过银行账号提供的存取款变动账户余额,而是直接修改账户余额属性。为了避免这种情况发生,可以采用 Swift 提供的访问控制机制。

```
39  let accountOfTom = PrivateAccount(owner: "Tom")
40  accountOfTom.deposit(amount: 10000)
41  accountOfTom.withdraw(amount: 8000)
42  accountOfTom.description()
43
44  accountOfTom.balance = 10000
45  accountOfTom.description()
```

```
deposit: 10000.0, balance: 10000.0
withdraw: 8000.0, balance: 2000.0
Tom's Private Account: EA716524-C699-4221-A07B-AFC0D4DD3F79, balance is 2000.0
Tom's Private Account: EA716524-C699-4221-A07B-AFC0D4DD3F79, balance is 10000.0
```

图 18.3　存取款的应用场景

如图 18.4 所示,将属性 balance 的访问权限设置为 private(set),即除了类本身的方法外,不能直接写属性。

```swift
class PrivateAccount: BankAccount {
    var accountID: String
    var owner: String
    private(set) var balance: Double
    func deposit(amount: Double) {
        balance += amount
        print("deposit: \(amount), balance: \(balance)")
    }
}
```

图 18.4　设置属性的访问权限

如图 18.5 所示,原来通过直接写属性的语句所在行出现编译器错误提示"Cannot assign to property：'balance' setter is inaccessible(不能直接对该属性赋值,balance 的写入是不可访问的)"。通过对余额属性的访问控制,防止了恶意修改敏感数据的行为。

```
let accountOfTom = PrivateAccount(owner: "Tom")
accountOfTom.deposit(amount: 10000)
accountOfTom.withdraw(amount: 8000)
accountOfTom.description()

accountOfTom.balance = 10000    ⊘ Cannot assign to property: 'balance' setter is inaccessible
accountOfTom.description()
```

图 18.5　提示私有属性不能直接赋值

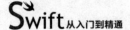

练习题

定义一个学生类 Student，含有 3 个属性：学号 id、年龄 age、姓名 name。定义一个自定义构造器。定义方法 changeID()，通过该方法可以修改学号。定义方法 description()，负责打印学生的基本信息，包括姓名、年龄和学号。要求学生的学号只能通过方法 changeID() 实现，不能直接通过实例的属性修改。定义一个学生实例，并进行初始化。打印该学生的基本信息，然后修改学生年龄和学号，重新打印学生的基本信息。

本章知识点

1. 介绍了访问级别的基本概念和主要的 3 个访问级别，包括 public、internal、private。

2. 应用访问控制的方法。

参考代码

练习题的参考代码如图 18.6 所示。

```
3   class Student {
4       private(set) var id: String
5       var age: Int
6       var name: String
7       init(age: Int, name: String) {
8           self.age = age
9           self.name = name
10          id = UUID().uuidString
11      }
12      func changeID(){
13          id = UUID().uuidString
14          print("Student's ID has been changed to \(id)")
15      }
16      func description(){
17          print("\(name)'s age is \(age) and id is \(id)")
18      }
19  }
20
```

图 18.6　练习题的参考代码

```
21  let studentTom = Student(age: 19, name: "Tom")
22  studentTom.description()
23
24  studentTom.age = 20
25  studentTom.changeID()
26  studentTom.description()
```

```
Tom's age is 19 and id is 0FF8CBB4-9E5E-432D-BABB-9B6DD0299898
Student's ID has been changed to BDC6BD7B-C945-4B68-BAC7-958FC422B271
Tom's age is 20 and id is BDC6BD7B-C945-4B68-BAC7-958FC422B271
```

图 18.6 （续）

第 19 章 类型操作符

Swift 中提供了两种类型操作符,分别为 is 和 as。通过 is 可以检查值的类型,通过 as 可以转换值的类型。本章将以类作为类型操作符的对象讲解这两个类型操作符的用法。

首先定义 3 个类,如图 19.1 所示,定义了一个 Student 类,这在前面章节已经出现过多次,这里对其成员进行裁剪。裁剪后,类 Student 只包含 3 个属性,分别为 name、age、id 以及一个构造方法。该构造方法需要传入 3 个值,分别为 Student 类中的 3 个属性进行初始化。

```
class Student {
    var name : String
    var age : Int
    var id : String
    init(name: String, age : Int, id : String) {
        self.name = name
        self.age = age
        self.id = id
    }
}
```

图 19.1 Student 类

如图 19.2 所示,定义了类 UnderGraduate,为 Student 类的子类。该类含有一个新属性 classDirector 和一个构造方法。它的构造方法引用父类的构造方法初始化从父类继承的属性 name、age、id。

如图 19.3 所示,定义的类 Graduate,也是 Student 类的子类。该类含有两个新属性 supervisor 和 researchTopic,以及一个构造方法。它的构造方法也引用了父类的

```
class UnderGraduate: Student {
    var classDirector : String
    init(name : String, age : Int, id : String,
        classDirector : String){
        self.classDirector = classDirector
        super.init(name: name, age: age, id: id)
    }
}
```

图 19.2　UnderGraduate 类

构造方法。

```
class Graduate: Student {
    var supervisor : String
    var researchTopic : String
    init(name : String, age : Int, id : String,
        supervisor : String, researchTopic :
        String){
        self.supervisor = supervisor
        self.researchTopic = researchTopic
        super.init(name: name, age: age, id: id)
    }
}
```

图 19.3　Graduate 类

19.1　类型检查

类型检查操作符 is 用来检查一个实例是否属于特定的类型。如果属于,则返回布尔值 true,否则返回 false。

如图 19.4 所示,在上例的基础上定义一个 Student 类的数组 students,其中包含 3 个元素,一个 Graduate 类实例,两个 UnderGraduate 类实例。数组 students 的类型由其成员元素的类型推断得到。由于数组 students 中实例的类型分别为 Graduate 类和 UnderGraduate 类,而这两个类又都是 Student 类的子类,所以数组 students 的类型为 Student 类。但是,在数组中所有元素又是按照各自的类型存储的,即一个 Graduate 类实例和两个 UnderGraduate 类实例。只是从 students 数组中获取元素

```
let students = [Graduate(name: "Sam", age: 23,
    id: "SY060115", supervisor: "Tim",
    researchTopic: "BPM"),UnderGraduate(name:
    "Jim", age: 19, id: "37060112",
    classDirector: "Lee"),UnderGraduate(name:
    "Kelly", age: 20, id: "37060113",
    classDirector: "Lee")]
```

图 19.4　类型检查

时，将统一作为 Student 类的实例处理。

如图 19.5 所示，继续定义两个变量 underGraduateNum 和 graduateNum 保存数组 students 中的 UnderGraduate 和 Graduate 的实例数。然后，通过一个 for-in 循环计算这两个变量的值。从程序的执行结果可以看出，数组 students 中的元素是按照各自创建时的类型保存的。

```
var underGraduateNum = 0
var graduateNum = 0
for item in students {
    if item is UnderGraduate {
        ++underGraduateNum
    } else if item is Graduate {
        ++graduateNum
    }
}
print("Undergraduate num is \(underGraduateNum) and graduate num is \(graduateNum)")
```

Undergraduate num is 2 and graduate num is 1

图 19.5　类型检查实例

19.2　类型转换

类型转换操作符 as 用来将一个实例向下转换为它的子类型。转换的过程可能成功，也可能失败，对于这种情况，类型转换操作符有两种形式，分别为"as?"和"as!"。"as?"返回一个可选值，"as!"则是强制向下转型并强制拆包。

当向下转型不确定是否能成功时，我们使用条件转换"as?"。条件转换返回一个可选值，如果向下转换失败，则返回 nil。

当向下转型确定可以成功时，我们使用强制转换"as!"。如果强制转换失败，则会触发运行时的异常。

students 数组中存在两种类型的实例，分别为 UnderGraduate 类和 Graduate 类。这两个类都是 Student 类的子类，但是又各有自己的扩展属性。在数组 students 中，所有实例都是按照各自创建时的类型保存的。当要将这些实例从数组中取出时，都会按照 Student 类取出。为了能够读出这两类实例中的不同属性，就需要把 students 数组中的实例转换成不同的类型。由于转换可能成功，也可能失败，所以需要使用"as?"条件转换符。

如图 19.6 所示,通过 for-in 循环取出数组 students 中的元素。首先通过条件转换符将其转换为 UnderGraduate 类型,如果成功,则打印 UnderGraduate 类中的属性 classDirector。如果失败,再尝试将其转换为 Graduate 类型,如果成功,则打印 Graduate 类中的属性 supervisor。如果失败,则继续下一个循环。通过执行结果可以看出,在这里使用"as?"条件转换符是恰当的,达到了预期的要求。

```
for item in students {
    if let theUnderGraduate = item as?
        UnderGraduate {
        print("\(theUnderGraduate.name)'s class
            director is \(theUnderGraduate.
            classDirector)")
    } else if let theGraduate = item as? Graduate {
        print("\(theGraduate.name)'s supervisor is
            \(theGraduate.supervisor)")
    }
}
```

```
Sam's supervisor is Tim
Jim's class director is Lee
Kelly's class director is Lee
```

图 19.6　类型转换实例

另外,Swift 提供了两种特殊的类型表示不确定类型,分别为 AnyObject 和 Any。其中,AnyObject 表示任何 class 类型的实例。Any 表示任何类型。可以使用强制转换"as!"的方式将 AnyObject 转换为一个明确的类型。

本章知识点

介绍了两种类型操作符 is 和 as。通过 is 检查值的类型,通过 as 转换值的类型。最后,通过实例讲解了这两种类型操作符的用法。

第 20 章 扩展

扩展就是向一个已有的类、结构体、枚举型或者协议类型增加新的功能。在 Swift 中，扩展包括：扩展计算型属性、扩展构造器、扩展方法、扩展下标。

声明一个扩展的格式如下：

```
extension OldType {
new functions
}
```

在扩展一个老的类型时要用关键字 extension。在花括号内定义新的扩展功能。可以扩展一个已有的类型，也可以使一个类型符合一个或者多个协议。对于前一种情况，上面已经介绍了扩展的格式。对于后一种情况，采用如下格式定义：

```
extension OldType: Protocol1, Protocol2 {
implement protocol's interface
}
```

除了要标示关键字 extension 外，还要在老的类型后加上扩展的具体协议名。另外，在花括号内要给出协议中接口的具体实现。

20.1 扩展计算型属性

本节通过一个例子说明如何扩展计算型属性。如图 20.1 所示，扩展了 Int 类型，添加了两个属性，用来表示不同的重量单位。其中，扩展定义中，用 self 表示类型本身的值，通过点运算符可以计算出变量值。变量 aChick 赋值等于 2kg，表示小鸡的质量为 2 千克，根据扩展属性，可以计算得到 aChick 的实际值为 2000，这里换算成了克。而 aElephant 赋值等于 8.ton，表示大象的质量为 8 吨，最后计算得到的值为 8 000 000，单位为克。

```
extension Int {
    var kg : Int {return self * 1000 }
    var ton : Int {return self * 1000000}
}
var aChick = 2.kg
```
2,000

```
var aElephant = 8.ton
```
8,000,000

图 20.1　扩展计算型属性

20.2　扩展构造器

本节通过一个例子说明如何扩展构造器。如图 20.2 所示，定义了一个结构体 Student，含 3 个属性 name、id、age，以及一个构造器。该构造器不需要任何输入参数，

```
struct Student {
    var name : String
    var id : String
    var age : Int
    init(){
        name = ""
        id = ""
        age = 0
    }
}

extension Student {
    init(name : String, id : String, age : Int){
        self.name = name
        self.id = id
        self.age = age
    }
}

let theStudent = Student(name: "Jack", id:
    "37060115", age: 18)
```

name
id
age 18

图 20.2　扩展构造器

直接对 3 个属性进行初始化。当需要在构造结构体实例时提供具体的属性值时,可以扩展这个结构体的构造器,提供一个含有 3 个参数的构造器,并在构造器中通过参数值为属性初始化。在常量实例 theStudent 定义的时候,可以选择调用该结构体扩展的构造器初始化实例。

20.3 扩展方法

本节通过一个例子说明如何扩展方法。如图 20.3 所示,在上例的基础上继续为结构体 Student 扩展一个方法 description(),用来将实例的基本信息打印出来。

```
extension Student {
    func description(){
        print("\(name)'s age is \(age) and id is
            \(id)")
    }
}
theStudent.description()
```

```
name "Jack"
id "37060115"
age 18
```

图 20.3　扩展方法实例

20.4 扩展下标

本节通过一个例子说明如何扩展下标。如图 20.4 所示,在上例的基础上又定义了一个班级 Class 的结构体,其中只有一个属性,即数组 students,数组的元素为 Student 类型。扩展结构体 Class,添加一个下标方法,根据传入的索引值读取 students 数组中相应元素的 name 属性。首先,创建一个 Class 结构体实例 theClass,然后向其 students 数组中添加 3 个 Student 结构体类元素。最后,常量 OneStudentName 通过下标方法获取 theClass 实例中数组 students 中第一个元素的 name 属性。

第 20 章　扩展

```
struct Class {
    var students = [Student]()
}

extension Class {
    subscript(Index : Int) -> String{
        return students[Index].name
    }
}

var theClass = Class()
theClass.students.append(Student(name: "Sam", id: "0",
    age: 16))
theClass.students.append(Student(name: "Tom", id: "1",
    age: 18))
theClass.students.append(Student(name: "Jeff", id: "2",
    age: 19))
let oneStudentName = theClass[0]
```
Sam

图 20.4　扩展下标实例

本章知识点

　　介绍了 4 种类型的扩展应用，包括：计算型属性扩展应用、构造器扩展应用、方法扩展应用、下标扩展应用。

第 21 章 内存管理

21.1 工作原理

Swift 使用自动引用计数(ARC)的机制解决应用程序的内存管理问题。一般来说,应用不需要进行内存管理,系统会自动管理类的实例,并在适当的时候释放其占用的内存。为了更好地利用 ARC 的机制,下面讲解其工作原理,帮助读者更好地理解这个过程,从而更好地为我所用。

当创建一个类的实例时,系统会分配一大片内存存储实例的相关数据。当该实例将不再被使用时,ARC 会自动释放实例占用的内存空间,从而使该内存空间可以被其他实例使用。此时,实例的相关数据,包括方法和属性,将不能再被访问和调用。当该实例在使用中,ARC 将通过跟踪和计算所有实例的引用数(即该实例被多少属性、常量及变量引用)确保该实例的内存空间不会被释放掉。当引用数不为 0 时,这个实例都不会被销毁。我们称属性、常量及变量对实例的引用为强引用。

图 21.1 定义了一个 Student 类,该类中定义了两个变量,分别为 name 和 age。另外,还定义了该类的构造器和析构器。构造器根据传入的参数值对类中的两个变量进行初始化,并打印信息表示该构造器被调用。析构器只有一条打印语句,用来表示该析构器被调用。定义两个 Student 类的可选型变量,分别为 studentA 和 studentB。创建一个 Student 类的实例,并将其赋值给 studentA,此时该实例被变量 studentA 引用。再将 studentA 赋值给变量 studentB,此时 studentB 也引用该实例,因此该实例的引用数为 2。然后,将 nil 赋值给 studentA,此时 studentA 对实例的引用失效。此时,只有变量 studentB 引用该实例,该实例的引用数为 1。注意观察控制台输出,此时

并没有打印析构器的信息"deinit is called"。再将 nil 赋值给变量 studentB，此时控制台输出"deinit is called"，说明实例的析构器被调用。此时，该实例的引用数为 0，说明该实例已无引用，故 ARC 自动销毁它。

```
class Student {
    var name : String
    var age : Int
    init(name : String, age : Int) {
        self.name = name
        self.age = age
        print("init is called")            "init is called\n"
    }
    deinit {
        print("deinit is called")          "deinit is called\n"
    }
}

var studentA : Student?                    nil
var studentB : Student?                    nil

studentA = Student(name: "Sam", age:       Student
    19)
studentB = studentA                        Student
studentA = nil                             nil
studentB = nil                             nil
```

图 21.1　Student 类及其实例

21.2　强引用循环

前面介绍的 ARC 机制可以帮助我们将不再被引用的实例自动销毁掉。通过判断一个实例被引用次数是否为 0 决定是否自动销毁实例。然而，有时会出现实例已经不会被使用了，但引用次数不为 0，这种情况就会导致内存泄漏，我们称这种情况为强引用循环。

图 21.2 定义了类 Student。该类有 3 个属性，分别为 name、age、supervisor，其中 supervisor 为可选属性。一个学生可能有导师，也可能没有，但是肯定有名字和年龄。另外，该类定义了构造器和析构器。构造器通过参数值为属性 name 和 age 初始化。析构器含有一条打印语句，当析构器被调用时，会自动打印相应的信息。

图 21.3 定义了类 Supervisor。该类含有两个属性，分别为 name 和 student。其中，student 属性为可选型，因为导师是一种带学生的资格，可能正带着学生，也可能暂时没有带学生。另外，该类也定义了构造器，用来初始化属性 name，定义了析构器，同

样为了在析构器被调用时能打印相应的信息。

```
class Student {
    var name : String
    var age : Int
    var supervisor : Supervisor?
    init(name : String, age : Int) {
        self.name = name
        self.age = age
        print("Student Class has been
            initialized.")
    }
    deinit {
        print("Student has been
            deinitialized!")
    }
}
```

图 21.2　Student 类

```
class Supervisor {
    var name : String
    var student : Student?
    init(name : String) {
        self.name = name
        print("Supervisor has been
            initialized.")
    }
    deinit{
        print("Supervisor has been
            deinitialized!")
    }
}
```

图 21.3　Supervisor 类

图 21.4 定义了两个变量，其中 theStudent 为类 Student 的可选型变量，theSupervisor 为类 Supervisor 的可选型变量，此时两者的值均为 nil。调用 Student 类的构造器为变量 theStudent 初始化，调用 Supervisor 类的构造器为变量 theSupvervisor 初始化，此时 Student 实例的引用数为 1，引用者为 theStudent，Supervisor 实例的引用数也为 1，引用者为 theSupervisor。然后建立学生 theStudent 和导师 theSupervisor 的师生关系，即将 theSupervisor 赋值给 theStudent 的 supervisor 属性，将 theStudent 赋值给 theSupervisor 的 student 属性。此时，Student 实例的引用数为 2，引用者为 theStudent 和 theSupervisor 的 student 属性。Supervisor 实例的引用数也为 2，引用者为 theSupervisor 和 theStudent 的 supervisor 属性。假设进行了一系列实例操作后，实例使用完毕，需要释放实例所占的存储空间。将变量 theStudent 赋值为 nil。此时，Student 实例的引用数减为 1，引用者为

```
var theStudent : Student?
var theSupervisor : Supervisor?

theStudent = Student(name: "Tom",
    age: 28)
theSupervisor = Supervisor(name:
    "Ian")

theStudent!.supervisor =
    theSupervisor
theSupervisor!.student = theStudent

theStudent = nil
theSupervisor = nil
```

图 21.4　强引用循环实例

theSupervisor 的 student 属性。接着将变量 theSupervisor 赋值为 nil。此时，Supervisor 实例的引用数减为 1，引用者为 Student 实例的 supervisor 属性。由于 Student 实例和 Supervisor 实例的引用数都为 1，所以 ARC 不会自动释放两者的内存空间，因为两个实例发生了循环引用。这种情况就会造成内存泄漏。

为了解决强引用循环的问题，Swift 提供了两种新的引用：弱引用和无主引用。循环引用中的一个实例引用另一个实例，可以用弱引用或无主引用替代原来的强引用，从而避免循环引用。对于生命周期中可能变成 nil 的实例采用弱引用。对于初始化后再也不会被赋值为 nil 的实例采用无主引用。

弱引用不会阻止 ARC 销毁被引用的实例，从而避免了循环强引用的问题。在声明一个属性或变量为弱引用时，我们使用关键字 weak。

如图 21.5 所示，为了解决上例的强引用循环，只做一点改动即可。修改类 Supervisor 的可选型属性 student，为其增加一个 weak 关键字即可。这里，由于类 Supervisor 的可选型属性 student 为弱引用，所以创建完实例 theStudent 和

```
class Supervisor {
    var name : String
    weak var student : Student?
    init(name : String) {
        self.name = name
        print("Supervisor has been
            initialized.")
    }
    deinit{
        print("Supervisor has been
            deinitialized!")
    }
}
```

图 21.5　强引用循环解决方案

theSupervisor，并建立两者的师生关系后，实例 Student 的引用数为 2，强引用者为 theStudent，弱引用者为 theSupervisor。实例 Supervisor 的引用数为 2，强引用者为 theSupervisor 和 theStudent 的 supervisor 属性。当实例使用完毕后，先将 theStudent 赋值为 nil，则实例 Student 的引用数变为 0，因为强引用者赋值为 nil，而弱引用者不影响 ARC 销毁实例，所以实例 Student 调用析构器，释放实例的内存空间。此时，实例 Supervisor 的引用数变为 1，因为强引用者 theStudent 已经被销毁了。将 theSupervisor 赋值为 nil，则引用数为 0。本例中，由于为 Supervisor 类的可选型属性 student 标为弱引用，所以不会出现强应用循环。同样，也可以通过修改 Student 类的可选属性 supervisor，增加 weak 关键字，达到解决强引用循环的问题。

无主引用和弱引用类似，也不会影响 ARC 释放内存。弱引用是可能有值或无值，而无主引用是永远有值。所以，无主引用用来定义非可选型。无主引用的关键字为 unowned，在声明的属性或变量前添加该关键字，表示该属性或变量为无主引用。

图 21.6 定义了 Person 类，该类含有两个属性，分别为名字 name 和汽车 car。一个人可能有车，也可能没有车，所以属性 car 为可选型。该类还定义了构造器，用来初始化属性 name，定义了析构器，打印析构器被调用的信息。

图 21.7 定义了类 Car，该类含有两个属性，分别为车主 owner 和车牌号 carNo。其中，车主为 Person 类的变量 owner。这种情况就会发生强引用循环。由于车肯定有车主，所以 owner 不可能为空，那么就不能用弱引用解决强引用循环的问题。在这种属性不可能为空的情况下，可以通过将其定义为无主引用解决强引用循环的问题。

```
class Person {
    var name : String
    var car : Car?
    init(name : String){
        self.name = name
    }
    deinit{
        print("person instance is
            destroyed")
    }
}
```

图 21.6　Person 类

```
class Car {
    unowned var owner : Person
    var carNo : String
    init(carNo : String, owner :
        Person) {
        self.carNo = carNo
        self.owner = owner
    }
    deinit{
        print("car instance is
            destroyed")
    }
}
```

图 21.7　Car 类

图 21.8 定义了变量 thePerson，该变量为可选型 Person 类。然后创建一个 Person 类实例并赋值给变量 thePerson，此时 Person 实例有一个强引用，引用者为 thePerson。接着，创建一个 Car 类的实例，该实例的 owner 属性赋值为 thePerson!，再将该实例赋值给 thePerson 的属性 car。此时，

```
var thePerson : Person?
thePerson = Person(name: "sam")
thePerson!.car = Car(carNo:
    "FK0151F", owner: thePerson!)
thePerson = nil
```

图 21.8　用无主引用解决强引用循环问题

Person 实例的引用数为 2，一个强引用来自 thePerson，一个无主引用来自 Car 实例。Car 实例的引用数为 1，是一个强引用，来自 thePerson。当这些实例使用完毕后，要回收资源，此时只将变量 thePerson 赋值为 nil 即可。当 thePerson 赋值为 nil 后，Person 实例的强引用就变为 0，此时只剩下一个 Car 实例的无主引用，而无主引用不会影响 ARC 释放资源，所以 thePerson 所占内存空间被释放。同时，由于 Person 实例被销毁，从而 Car 实例的引用数变为 0，所以 Car 实例也会被 ARC 销毁。最终通过无主引用的方式圆满地解决了强引用循环问题。

前面两个例子分别介绍通过弱引用和无主引用解决强引用循环的问题。这两种

引用适用于不同的场景。当两个类中的两个属性都为可选型时，产生的循环强引用可以通过弱引用的方式解决。当两个类中的属性一个为可选型，一个为非可选型，这种情况下产生的循环强引用可以通过无主引用解决。

除上述两种情况外，还有第三种情况，就是两个类中的两个属性都必须有值，初始化后就不会为空。在这个场景下也会产生强引用循环。那么，解决这种情况的方法就是一个类中的属性使用无主属性，另一个类中的属性使用隐式解析可选属性。

假定上例的场景变为：每个人都肯定有车，同时每个车都有车主。如果变成这个场景，就需要修改上面的类定义了。如图 21.9 所示，修改类 Person 的定义，将属性 car 修改为 Car 类型隐式解析可选型。同时要修改构造器的定义，在构造器中必须为 car 属性赋值。这里，通过调用 Car 类的构造器创建实例，而 Car 类的构造器需要传入一个 Person 类的实例作为参数传入，这个参数值应该为 self。此时，Person 类的实例还没有创建完，因为 car 属性还没有值。这就用到了隐式解析可选型，前面将属性 car 定义为隐式解析可选型，所以默认值就是 nil。当 Person 类中的 name 属性被赋值，self 就有值了。所以，用 self 作为参数传给 Car 的构造器在 Person 类的构造器里给 self 的属性赋值就没有问题了，也不会产生强引用循环。另外，Car 类的定义不需要调整，仍然是将其 owner 属性标识为无主引用。

如图 21.10 所示，先创建 Person 实例并赋值给变量 thePerson。然后查看 thePerson.car.carNo 和 thePerson.car.owner.name 属性，可以看出通过无主引用和隐式解析可选属性的方法，不仅可以解决循环引用的问题，而且不需要显式解析可选属性，运行结果正确。

```
class Person {
    var name : String
    var car : Car!
    init(name : String, carNo :
        String){
        self.name = name
        self.car = Car(carNo: carNo,
            owner: self)
    }
    deinit{
        print("person instance is
            destroyed")
    }
}
```

图 21.9　修改后的 Person 类

```
var thePerson = Person(name: "Tom",
    carNo: "FK0515H")
thePerson.car.carNo
thePerson.car.owner.name
```

Person
"FK0515H"
"Tom"

图 21.10　无主引用实例

21.3 闭包中的强引用循环

21.2 节介绍了两个类的实例属性互相保持对方的强引用而导致强应用循环。本节将介绍发生在类实例和闭包之间的强引用循环及其解决方法。由于类和闭包都是引用类型，当将一个闭包赋值给一个类实例的某个属性时，如果闭包体中引用了该类实例（属性或方法），那么就会导致闭包捕获自己，从而产生强引用循环。

图 21.11 定义了类 Person，包含 3 个属性名字 name 和汽车品牌 carBrand，还有一个闭包类型的延迟变量 description，该闭包没有参数，返回值为字符串，默认值为打印车主名字和汽车品牌信息。另外，还有构造器和析构器。

```
class Person {
    var name : String
    var carBrand : String

    lazy var description : Void ->
        String = {
        return "\(self.name) has a \
            (self.carBrand) car"
    }

    init(name : String, carBrand :
        String){
        self.name = name
        self.carBrand = carBrand
    }
    deinit{
        print("person instance is
            destroyed")
    }
}
```

图 21.11　Person 类的定义

如图 21.12 所示，调用构造器创建 Person 类实例，并赋值给可选型变量 thePerson。此时，Person 实例有两个强引用，引用者为 thePerson 和默认的闭包，该闭包有一个强引用，引用者为 Person 实例。然后调用变量 thePerson 中的闭包属性 description，系统正确打印出了车主和汽车品牌信息。实例使用完毕后，将变量 thePerson 赋值为 nil。此时，Person 实例的强引用为一个，为默认的闭包。同时，闭包保持原来的强引用，从而造成强引用循环。由于强引用循环，所以系统没有自动调用 Person 实例的析构器，ARC 未能成功释放相关的内存空间。

为了解决这个问题，Swift 提供了一种叫闭包捕获列表的方法，即在定义闭包的

```
var thePerson : Person? = Person(name:
    "Tommy", carBrand: "Nissan")
thePerson!.description()
thePerson = nil
```

```
Person
"Tommy has a Nissan car"
nil
```

图 21.12　闭包中的强引用循环实例

同时定义捕获列表作为闭包定义的一部分。捕获列表定义了闭包体内一个或多个引用类型为弱引用或者无主引用。捕获列表中的每项都由一对元素组成，即关键字（weak 或 unowned）和类实例的引用（self 或 delegate＝self.delegate!）。捕获列表放在闭包的参数列表和返回类型的前面。如果闭包没有明显的参数列表或者返回类型，则可以把捕获列表和关键字 in 放在闭包的最前面。

当闭包和捕获的实例是相互引用，用完就同时销毁，应将闭包内的捕获定义为无主引用。如果捕获的实例可能为 nil，应将闭包内的捕获定义为弱引用。

前一个例子中造成强引用循环的原因是闭包和捕获的实例总是互相引用，同时销毁，所以应该使用无主引用解决这个问题。如图 21.13 所示，只需要在上例的基础上为闭包变量 description 增加一行代码，即在闭包体的最前面部分加上捕获列表。同样的执行代码，在将 nil 赋值给 thePerson 时，系统显示 Person 实例调用了析构器，进行了销毁。

```
class Person {
    var name : String
    var carBrand : String

    lazy var description : Void ->
        String = { [unowned self] in
        return "\(self.name) has a \
            (self.carBrand) car"
    }

    init(name : String, carBrand :
        String){
        self.name = name
        self.carBrand = carBrand
    }
    deinit{
        print("person instance is
            destroyed")
    }
}

var thePerson : Person? = Person(name:
    "Tommy", carBrand: "Nissan")
thePerson!.description()
thePerson = nil
```

```
"person instance is destroyed\n"

Person

"Tommy has a Nissan car"
nil
```

图 21.13　闭包捕获列表实例

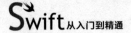

本章知识点

1. 介绍了自动引用计数（ARC）的工作原理。

2. 介绍了如何通过弱引用和无主引用解决实例中属性互相引用造成的强引用循环。

3. 介绍了通过闭包捕获列表解决类实例与类中闭包之间的强引用循环的方法。

第 22 章 高级运算符

第 1 篇的语法基础介绍了各种 Swift 基本运算符，本章将介绍高级运算符，内容包括：位运算符、溢出运算符及运算符函数。

22.1 位运算符

位运算符指对一个数据中的每个位进行操作。位运算符在各种底层开发中有广泛的应用。本节将介绍：按位取反运算符、按位与运算符、按位或运算符、按位异或运算符、按位左移运算符和按位右移运算符。

按位取反运算符的操作符为"~"，它能将一个数按照二进制位进行取反。如图 22.1 所示，定义一个变量 origin 为无符号 8 位整型，也就是说，变量 origin 是一个

图 22.1 位运算符

整数，取值范围是 0～255，这里指的是十进制范围，二进制范围是 00000000～11111111，无符号指没有符号位，8 位指的是二进制的位数。origin 的初值为 15，用二进制表示为 0b00001111。将 origin 按位取反，即 00001111 中的每一位取反，为 0 的变为 1，为 1 的变为 0，结果赋值为变量 result。此时 result 的值为 240，用二进制表示就是 11110000。这里需要注意的是，origin 的赋值无论为二进制的形式 0b00001111，还是十进制的形式 15，对取反的结果是没有影响的。如果将 origin 赋值为 240，即二进制的 0b11110000，对其取反后的二进制结果为 0b00001111，换算成十进制就是 15。

二进制和十进制只是一个数的表现形式，在系统中实际是保存为二进制形式的，十进制只是比较直观而已。

按位与运算符为"&"，它将两个数的二进制形式按照位进行"与"操作，然后将每个位的操作结果再组合成一个新的数。如图22.2所示，定义了两个操作数operatorA和operatorB，均为无符号8位整型，分别赋初值为二进制的11001100和10101010。将这两个数进行按位"与"操作，只有相应位均为1的时候，位结果才为1，运算后的结果为10001000，等价于十进制的136。

按位或运算符为"|"，它将两个数的二进制形式按照位进行"或"操作，然后将每个位的操作结果再组合成一个新的数。如图22.2所示，将operatorA和operatorB进行按位"或"操作，只有相应位均为0的时候，位结果才为0，运算后的结果为11101110，等价于十进制的238。

按位异或运算符为"^"，它将两个数的二进制形式按照位进行"异或"操作，然后将每个位的操作结果再组合成一个新的数。如图22.2所示，将operatorA和operatorB进行按位"异或"操作，只要相应位不相等的时候，位结果就为1，运算后的结果为01100110，等价于十进制的102。

按位左移运算符为"<<"，按位右移运算符为">>"，它对一个数的二进制形式按照指定位数进行左移或者右移。对一个数的左移或者右移，相当于对这个数进行乘以2或者除以2的操作。在对一个数进行左移或者右移时，超出存储位数的位会被丢弃，产生的空白位用0填充。图22.3定义了一个常量shift为无符号8位整型，赋初值为二进制的10101010，相当于十进制的170。将shift左移1位后，二进制值为01010100，最后一位用0填充，十进制值为84。将shift左移2位后，二进制值为10101000，最后两位用0填充，十进制值为168。将shift右移1位后，二进制值为01010101，开头的一位用0填充，十进制值为85。将shift右移2位后，二进制值为00101010，开头的两位用0填充，十进制值为42。

```
var operatorA : UInt8 = 0b11001100         204
var operatorB : UInt8 = 0b10101010         170

result = operatorA & operatorB             136
result = operatorA | operatorB             238
result = operatorA ^ operatorB             102
```

```
let shift : UInt8 = 0b10101010             170
result = shift << 1                         84
result = shift << 2                        168
result = shift >> 1                         85
result = shift >> 2                         42
```

图22.2　按位与、或、异或运算　　　　　　　　图22.3　按位左移和右移

22.2 溢出运算符

当一个整数的赋值超过它的最大限度时，系统会报错。例如，上例中定义的变量都是无符号8位整型，那么它的取值范围就是0～255。当为Int8类型的变量赋值为266时，系统就会报错，如图22.4所示。

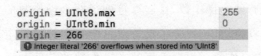

图22.4　溢出实例

Swift提供了3种溢出操作符，用来支持整数的溢出运算，即如果发生了上例中出现的溢出运算，系统不会报错，而会将溢出部分舍弃。溢出操作符包括：溢出加法"&+"、溢出减法"&-"、溢出乘法"&*"。

图22.5　溢出运算符

如图22.5所示，前面已经定义 result 为 UInt8 类型，那么 result 的取值范围就是0～255。UInt8.max 等于255，即二进制的11111111，UInt8.min 等于0，即二进制的00000000。UInt8.max 加1或者 UInt.min 减1都会导致溢出，因为从算术角度讲，UInt8.max 加1等于二进制的100000000，需要9位二进制数才能保存，而 result 最多只能保存8位。同样，UInt8.min 减1会发生向下溢出。如果 UInt8.max 溢出加1，则会截断最高位1，得到二进制的00000000。如果 UInt8.min 溢出减1，同样会截断最高位1，得到二进制的11111111。这种情况系统不会报错，因为在发生溢出运算的时候已经进行了处理。UInt8.max 乘以2后，等于二进制的111111110，系统会向上溢出，如果使用溢出乘，则会截断最高位，运算结果为11111110，即十进制的254。

22.3 运算符函数

Swift支持在类和结构体中对现有的操作符提供自定义的实现，这个过程我们称之为运算符重载。图22.6定义了一个结构体Point，用来表示一个坐标点，它含有两

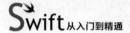

个变量 x 和 y,分别表示 x 坐标值和 y 坐标值。然后我们重载了现有的减号运算符,赋予它新的运算语义,即实现两个结构体 Point 类型的点的相减,具体来讲就是,两个点的相减就是两个点的 x 坐标和 y 坐标分别相减,得到的值分别作为一个新的点的 x 坐标和 y 坐标,这个新的点就是两个点相减的结果。我们创建了两个点(3,5)、(1,6),分别赋值给常量 a 和 b。然后将 a−b 的结果赋值给 c。从结果可以看出两个点相减的运算符执行结果正确。

上例中的运算符有两个操作数,且运算符处于两个操作数的中间,这种运算符我们称之为双目中缀运算符。Swift 除了支持双目中缀运算符的重载,也支持单目运算符的重载,即只有一个操作数。根据运算符在操作数的前面和后面,又分为前缀运算符和后缀运算符。重载一个前缀或后缀运算符时,要在声明运算符函数的时候在关键字 func 前面加上关键字 prefix 或 postfix。如图22.7所示,定义单目前缀运算符"−",用来取一个负点,即将一个点的 x 坐标和 y 坐标都取负得到的新的点。这里,我们创建一个点(3,4),赋值给常量 point,然后将−point 赋值给常量 pointN。从运行结果看出,pointN 的值为(−3,−4)。

```
struct Point {
    var x = 0
    var y = 0
}
func -(a: Point,b: Point)-> Point {
    return Point(x: a.x-b.x, y:
        a.y-b.y)
}
let a = Point(x: 3, y: 5)
let b = Point(x: 1, y: 6)
let c = a - b
```

```
x 2
y -1
```

图 22.6　结构体 Point

```
struct Point {
    var x = 0
    var y = 0
}
prefix func -(point: Point)-> Point{
    return Point(x: -point.x, y: -
        point.y)
}
let point = Point(x: 3, y: 4)
let pointN = -point
```

```
x -3
y -4
```

图 22.7　单目运算符的定义

赋值运算符"="与其他运算符结合起来形成的运算符称之为复合赋值运算符。如图 22.8 所示,定义了减法赋值运算符"−=",实现了两个点相减,并将结果赋值给被减数。这里需要注意的是,赋值运算符没有返回值,运算结果是通过参数的值被改变的方式实现的,为了将参数的值的变化保存到函数体外,需要将该参数定义为 inout 类型。这里创建了两个点,point 为(3,6),decrement 为(2,8)。point 减赋值

decrement 的结果为：point 被赋值为(1,−2)。

```
struct Point {
    var x = 0
    var y = 0
}

func -=(inout origin:Point, decrement:
    Point){
    origin.x = origin.x - decrement.x
    origin.y = origin.y - decrement.y
}

var point = Point(x: 3, y: 6)
let decrement = Point(x: 2, y: 8)
point-=decrement
```

```
x 1
y -2
```

图 22.8　减法赋值运算符的定义及实例

　　Swift 中没有提供判断类或结构体等价的操作符。可以通过重载等价操作符"=="或者不等价操作符"!="的方式实现对两个类或者结构体是否等价的运算。图 22.9 重载了双目中缀操作符"=="，该函数有两个参数，均为 Point 类型，返回值为布尔型。在函数体内，分别对两个点的 x 坐标和 y 坐标进行是否等价判断，得到的结果为布尔值，再将两个布尔值进行"与"运算。只有两个点的 x 坐标和 y 坐标都相等

```
struct Point {
    var x = 0
    var y = 0
}

func == (point1: Point, point2: Point)-
    >Bool{
    return (point1.x == point2.x) &&
        (point1.y==point2.y)
}

let point1 = Point(x: 1, y: 2)
let point2 = Point(x: 2, y: 3)
let point3 = Point(x: 1, y: 2)

if point1 == point2 {
    print("point1 is equal to point2")
}
if point1 == point3 {
    print("point1 is equal to point3")
}
```

```
point1 is equal to point3
```

图 22.9　等价运算符的定义及实例

的时候,返回值才会是 true。然后,创建了 3 个点,分别为:point1 等于(1,2),point2 等于(2,3),point3 等于(1,2)。通过重载的等价操作符可以直接判断两个点是否相等,根据运行结果可知,point1 等于 point3。

本章知识点

1. 介绍了按位取反运算符、按位与运算符、按位或运算符、按位异或运算符、按位左移和按位右移运算符及其应用。
2. 介绍了溢出加"&+"、溢出减"&-"、溢出乘"&*"及其应用。
3. 介绍了双目运算符、单目运算符、复合赋值运算符及其应用。

第 4 篇

应 用 篇

前面的章节主要介绍了 Swift 的语法，所有代码都是在 playground 中完成的。学习完 Swift 语法后，就具备开发苹果应用的基础了，后面的章节将从如何创建 iOS 工程入手，逐步介绍如何运用前面学习的知识构建一个苹果应用。由于本书专注于讲解如何应用 Swift 语言，所以前面一直都没有讲 iOS 开发技术，而且 iOS 开发技术涉及的内容比较多。本书最后一部分以应用 Swift 语言解决实际开发问题为目标，不展开讨论涉及 iOS 开发技术的部分内容，如果读者想进一步了解 iOS 开发技术，可参考相关技术文档，相信有了 Swift 的语言基础，在学习 iOS 开发技术时，读者一定会如鱼得水。

第 23 章 苹果应用

本章主要介绍构建一个苹果应用的过程和相关知识,帮助读者快速进入应用开发的状态中。

23.1 一个简单的应用

所有的苹果应用项目都是从创建一个工程开始的。如图 23.1 所示,在启动页面中选择"Create a new Xcode project"。

图 23.1 创建一个 Xcode 项目

如图 23.2 所示，进入新项目的模板选择页面。这个页面中列出了所有苹果应用的开发模板。开发模板分为四大类：iOS、watch OS、OS X 和 Other。iOS 主要是手机应用方面的。watch OS 为苹果手表的应用模板。OS X 为苹果计算机上的应用程序模板。其中 iOS 又分为应用模板和框架及库模板。这里选择 Application，右边会同步显示 Application 中的模板，具体有主从应用模板"Master-Detail Application"、基于页面模板"Page-Based Application"、单视窗模板"Single View Application"、多视图切换模板"Tabbed Application"。这里选择最简单的"Single View Application"，然后单击 Next 按钮。

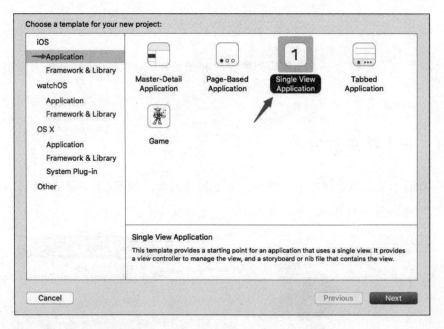

图 23.2 创建 Single View Application 工程

如图 23.3 所示，在项目信息表单中需要填写项目名称"Product Name"，这里填写 Calculator，即计算器，后面要逐步实现一个计算器。还需要填写：组织名"Organization Name"、组织标识"Organization identifier"、捆绑标识符"Bundle identifier"（它由产品名加上公司名）、开发语言"Language"（这里选择 Swift）、设备"Devices"（这里选择 iPhone）。设置完成后，单击 Next 按钮。

如图 23.4 所示，在弹出的界面里要选择项目保存的位置。建议读者创建好一个文件夹，将项目相关的文件放在里面，这样便于日后管理。然后单击创建 Create 按钮。

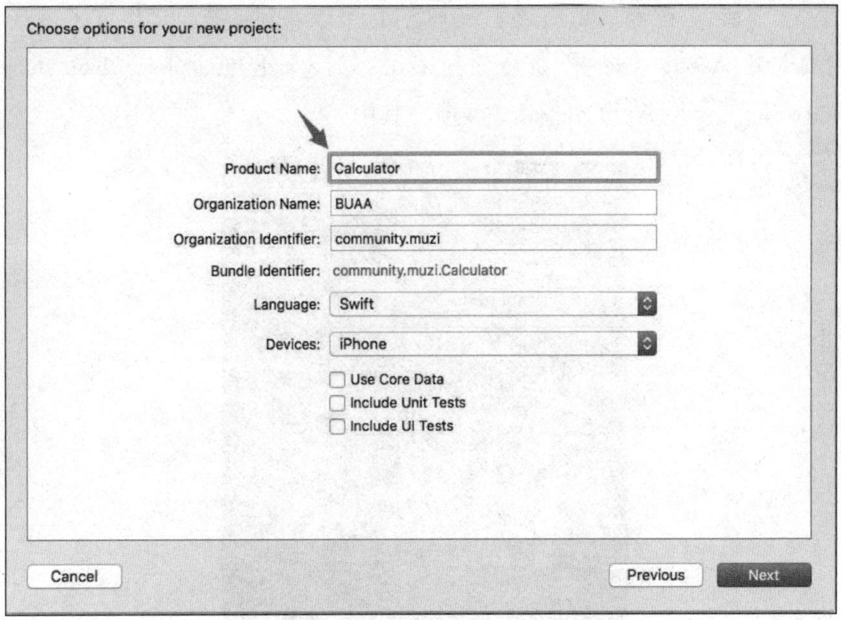

第 23 章　苹果应用　Chapter 23

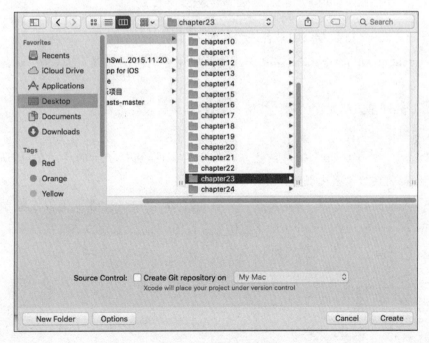

图 23.3　项目信息输入页面

图 23.4　文件保存目录

如图 23.5 所示，已经成功创建了一个项目，可以看到系统自动生成了一些通用的文件，主要包括 Assets.xcassets、LaunchScreen.storyboard、Info.plist、Main.storyboard、ViewController.swift、AppDelegate.swift。其中：

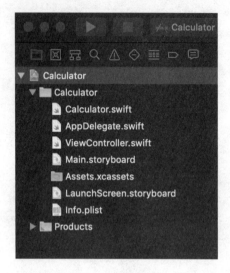

图 23.5　自动创建的文件列表

Assets.xcassets 是专门用来管理图片的文件夹。

LaunchScreen.storyboard 是系统创建的应用启动时的加载页面，是一个故事板文件。

Info.plist 是工程属性的描述性配置文件，如果要设置或修改工程的某些属性，可以直接编辑这个文件。

Main.storyboard 是系统默认建立的故事板文件，包括主要的界面部分，所有的界面工作都将在故事板中展开。

ViewController.swift 中定义了 ViewController 类，该类继承了视图控制器类 UIViewController。ViewController 类作为根视图，主要负责用户事件控制方面。

AppDelegate.swift 的代码如下：

```
import UIKit
@UIApplicationMain
class AppDelegate: UIResponder, UIApplicationDelegate {
var window: UIWindow?
```

```
func application (application: UIApplication, didFinishLaunchingWithOptions
launchOptions:[NSObject: AnyObject]?) ->Bool {
return true
    }
func applicationWillResignActive(application: UIApplication) {
    }
func applicationDidEnterBackground(application: UIApplication) {
    }
func applicationWillEnterForeground(application: UIApplication) {
    }
func applicationDidBecomeActive(application: UIApplication) {
    }
func applicationWillTerminate(application: UIApplication) {
    }
}
```

从上面的代码中可以看出，AppDelegate.swift 定义了应用程序委托对象，它继承了 UIResponder 类，并实现了 UIApplicationDelegate 委托协议。UIResponder 类使其子类 AppDelegate 具有处理相关事件的能力。UIApplicationDelegate 委托协议则使 AppDelegate 成为应用程序委托对象，从而能够响应应用程序的各个生命周期行为。AppDelegate 类继承的一系列方法在应用程序生命周期的不同阶段都会被回调。

如图 23.6 所示，选中 Main.storyboard，可以看到界面编辑页面出现在中间的编辑区。箭头标示的地方显示"wAny hAny"。假定是在 iPhone6 Plus 的设备上使用，则应该将其设置为"wCompact hRegular"，即宽度尺寸紧凑，高度尺寸正常，通过鼠标将矩形框调整为如图 23.7 所示即可。

这里使用到屏幕适配技术，官方文档称其为"Size Class"。在我们的工程中，默认已经开启了"Size Class"功能，如图 23.8 所示。如果没有开启，则需要手动开启。

Size Class 面板如图 23.7 所示，它是一个九宫格，即由 9 个方格组成，一共可以组合成 9 种布局方案，用来满足各种苹果设备的屏幕适配问题。这 9 个方案是通过两个维度的坐标值的组合形成的。这两个维度为宽度（w）和高度（h）。9 个布局方案分别为：

wCompact hCompact：用于 iPhone（3.5 英寸、4 英寸、4.7 英寸屏幕）横屏的情况。

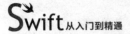

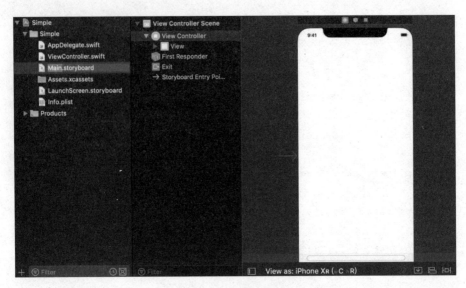

图 23.6 故事板页面

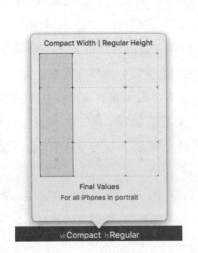

图 23.7 屏幕尺寸适配

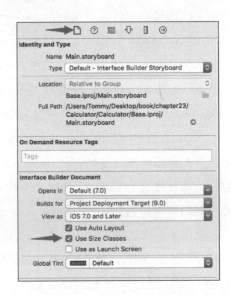

图 23.8 Size Classes 选项

wAny hCompact：用于 iPhone（所有尺寸）横屏的情况。

wRegular hCompact：用于 iPhone（5.5 英寸屏幕）横屏的情况。

wCompact hAny：用于 iPhone（3.5 英寸、4 英寸、4.7 英寸屏幕）竖屏的情况。

wAny hAny：用于所有情况。

wRegular hAny：用于 iPad 的横屏和竖屏的情况。

wCompact hRegular：用于 iPhone（所有尺寸）竖屏的情况。

wAny hRegular：用于 iPhone（所有尺寸）竖屏和 iPad 横屏、竖屏的情况。

wRegular hRegular：用于 iPad 横屏和竖屏的情况。

如图 23.9 所示，调整 Size Class 为"wCompact hRegular"后的界面，可以看到界面已经变成一个矩形，形状和 iPhone 手机相似。

如图 23.10 所示，在 Xcode 中的右下角位置选中对象库按钮，打开对象库，会看到一堆可视化组件。选中文本标签组件 Label，并将其拖曳到中间的设计界面中，尽量将其摆放在中间位置。摆放组件的时候，会看到蓝色虚线，帮助我们将组件摆放到中间对齐的位置。

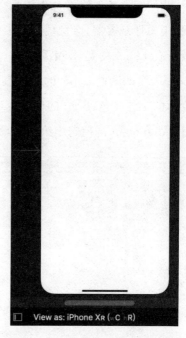

图 23.9　wCompact hRegular 效果界面

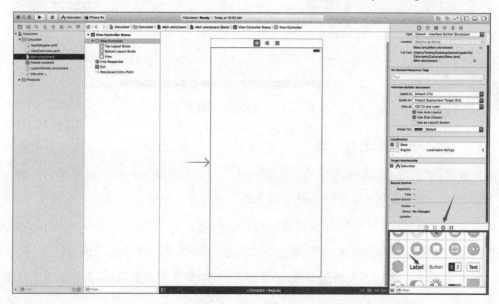

图 23.10　对象库

如图 23.11 所示，摆放好 Label 组件后，选中组件的边缘，然后拖动调整其大小。这里，我们将原始的组件大小放大一些。选中 Label 的情况下，注意右上角的导航部分，选中属性观察器，会发现打开了如图 23.10 所示的属性表单。在这个表单中，可以设置所有 Label 的显示相关属性。这里只调整几个属性：Text 的值设置为"It is calculator"，即设置 Label 的文本值；Color 的值通过下拉列表选中蓝色，即设置 Label 中文本的字体颜色；Font 的值设置为"System 30"，即设置 Label 中文本的字体类型和尺寸；Alignment 的值选中第二个，即设置 Label 中的文本居中显示。

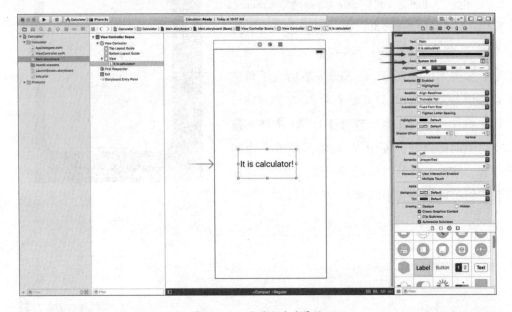

图 23.11　设置 Label 属性

设置完之后，就可以运行程序了。如图 23.12 所示，单击 Xcode 界面中左上角的运行按钮，就会启动模拟器 Simulator，然后打开一个模拟 iPhone6 Plus 手机的页面。

运行结果未能完整显示整个模拟的手机界面。为了能够在屏幕中完整地显示，可以通过设置 Simulator 中的显示比例解决这个问题。如图 23.13 所示，在 Simulator 中的菜单栏中选择 Window，然后选择 Scale，在子菜单中选择合适的缩放比例，这里选择的是 33%。然后再重新运行程序，发现整个模拟器的页面都可以在屏幕里显示了，如图 23.14 所示。

第 23 章 苹果应用

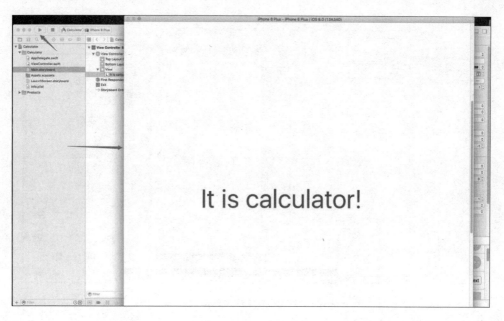

图 23.12 运行结果

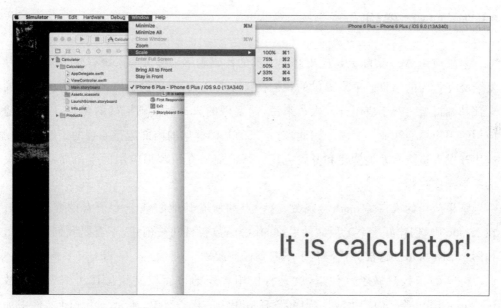

图 23.13 显示比例的调整

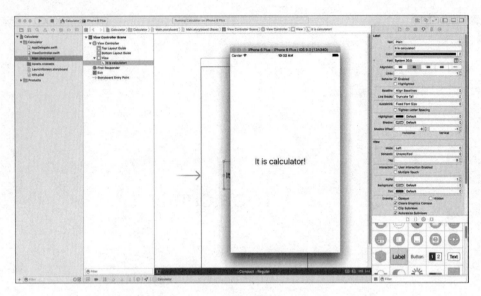

图 23.14　调整显示比例后的效果

23.2　MVC 架构

苹果应用 App 实际上是应用开发人员的代码和苹果提供的系统框架之间的一种复杂交互行为的结果。苹果提供的系统框架非常强大，它提供了应用开发人员需要的全部基础框架。对于应用开发人员来说，需要做的只是对基础框架的配置，以及构建应用的用户界面和业务逻辑。因此，对苹果的基础框架的内部运行原理的了解会帮助应用开发人员更有效地构建和开发应用。下面简要介绍苹果应用的生命周期及相关的系统操作过程。

苹果应用的入口是 main() 函数，这一点和所有其他的基于 C 语言的应用一样。在 Xcode 中，创建好一个工程文件后，main() 函数会自动被创建，不需要我们手动写，如图 23.15 所示。这部分代码几乎不需要任何改动。

main() 函数只做了一件事，就是将应用程序的控制权交给 UIKit 框架。函数 UIApplicationMain() 接手应用程序，创建应用中的核心对象、从 storyboard 文件中加载用户界面、调用开发人员编写的代码进行应用的初始化，并启动应用的生命周期管理，进入事件循环队列。

```
#import <UIKit/UIKit.h>
#import "AppDelegate.h"
int main(int argc, char * argv[])
{
    @autoreleasepool {
        return UIApplicationMain(argc, argv, nil,
NSStringFromClass([AppDelegate class]));
    }
}
```

图 23.15　系统自动创建的代码

在应用启动阶段，UIApplicationMain（）函数负责创建关键对象，并启动应用运行。苹果应用的核心是 UIApplication 对象，它负责系统和应用中的对象之间的交互行为。苹果的应用均采用了模型-视图-控制器（Model-View-Controller，MVC）架构，这个架构是一个经典的设计模式，它很好地将应用数据和业务逻辑从用户界面中剥离出来，如图 23.16 所示。

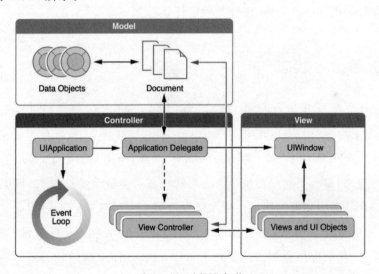

图 23.16　MVC 架构

图 23.16 描述了苹果应用的 MVC 架构中有哪些关键对象，以及它们之间的关系。下面简要介绍每个对象的功能。

UIApplication：该对象管理事件循环队列和应用的行为，并将应用的关键状态变迁和发生的特殊事件报告给它的代理，这个代理往往是应用开发人员定制的，用来处理这些事件。

Application Delegate：该对象是应用开发人员定制的核心代码。它和UIApplication对象一起负责处理应用的初始化、状态变迁的管理，以及应用事件的处理，它存在于所有应用中，所以常常负责创建应用的初始数据结构。

Document：该对象不是必需的，但是它可以提供一种便利的方式对数据进行分类。

Data Objects：该对象存储应用中的数据。

View Controller：该对象负责应用的呈现。它管理一个独立的视图以及该视图的子视图。UIViewController类是所有视图控制器的基类。它提供了所有视图类的默认功能：加载视图、呈现视图、旋转视图等。

UIWindow：该对象负责协调一个或多个视图在屏幕上的呈现过程。绝大多数应用只有一个呈现应用内容的主窗口，但是应用可以有一个显示内容的外部显示器的窗口。

Views and UI Objects：该类负责提供可视化的应用内容。它负责在指定的范围内绘制可视化的视图组件，并响应外界的事件。

23.3 应用运行状态

苹果应用总是处于5个状态中，即Not running（未运行）、Inactive（未激活）、Background（后台运行）、Active（激活）、Suspended（被挂起）。系统负责根据外界动作将应用从这5个状态中的一个切换到另一个。例如，当用户按下Home键后，或者有一个电话打进来，那么当前正在执行的应用的状态将会被改变。图23.17显示了苹果应用在这5个状态中切换的路径。应用还没有被启动或者被启动过但是又被系统终结了，那么此时就处于Not running状态。当应用正在前台运行的时候，当前没有接收到任何事件，那么应用会短暂停留在Inactive状态（即未激活状态），然后便转移到别的状态。该状态不能处理系统事件。当应用启动完毕后，进入Active，即激活状态，可以处理各种事件。这个状态是前台应用的最常见状态。Active和Inactive之间会根据用户对应用的操作情况进行切换。当应用处于Inactive时，如果长时间未被激活，会转入 Background，即后台运行状态。应用进入后台运行状态后依然可以执行代码，当执行完毕后或者无可执行代码时，将会进入 Suspended，即被挂起状态。大部分应用进入 Background 后很快会转到 Suspended 状态。处于被挂起状态时，应用不能执行代码。当系统内存紧张时，会将应用结束，并进入 Not running。

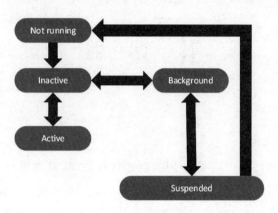

图 23.17　应用运行状态

通过应用中的代理委托机制，大部分的状态转移发生时，都会调用应用中的相关处理方法。通过这些方法，开发人员可以对应用的状态变化进行适当处理。下面列出了相关的方法：

application:willFinishLaunchingWithOptions：该方法在应用启动时执行。

application:didFinishLaunchingWithOptions：该方法在应用即将呈现给用户时执行，可以在此方法中进行最后的初始化工作。

applicationDidBecomeActive：该方法在应用即将变为前台 Active 状态时执行。

applicationWillResignActive：该方法在应用从前台的 Active 状态变化到 Inactive 状态时执行。

applicationDidEnterBackground：该方法在应用已经进入后台时执行，即处于 Background 状态，该状态随时有可能进入 Suspended 状态。

applicationWillEnterForeground：该方法在应用从后台转入前台时被执行，也就是从 Background 状态进入 Inactive 状态时被执行。

applicationWillTerminate：该方法在应用即将被中止（Not running 状态）时执行。如果应用被挂起（处于 Suspended 状态），则该方法不会被调用。

第 24 章
计 算 器

第 23 章简要介绍了一个苹果应用的创建过程,本章将在该应用的基础上实现一个计算器应用,功能和 iPhone 自带的计算器一致,主要包括:整数的加、减、乘、除四则运算,取百分比,取正负,清零等功能。这是一个极其简单的计算器,也是一个在 iPhone 中使用频率非常高的一个应用。通过这个例子,可以进一步了解我们平时很熟悉的应用是如何一步步构建出来的,同时还可以应用前面学到的 Swift 语言解决实际问题。关于如何构建一个空的应用项目,这里就不再介绍了,请读者参考第 23 章的内容。

24.1 界面设计

计算器的界面非常简单,只有一个操作面板。在这个操作面板上应该有 0,1,…,9 的数字键,用来输入整数操作数。还应该有小数点,用来输入小数部分,也就是浮点数。还应该有加、减、乘、除以及取正负和取百分数的操作运算符。还应该有一个等号,用来启动计算,得出结果。最后应该有一个清除屏幕的按键,用来快速归零。

这个界面虽然很简单,但是对于没有用户界面设计基础的人来说,设计出漂亮的界面是非常困难的。关于如何设计出美观大方,又有良好用户体验的苹果应用,这个内容已经超出本书的范畴。我们的目标是把注意力集中在如何构建一个实际的应用,如何用 Swift 编写相关的代码上。如图 24.1 所示,这是 iOS 自带的计算器。读者可以通过滑动苹果手机界面的最低端将其启动。打开苹果的计算器应用后,通过同时按

第 24 章　计算器

HOME 键和锁屏键可以实现截屏，将苹果计算器的界面截取下来。此时需要使用图片处理软件做一些处理，首先去掉初始值 0，该值的输出由按键产生。然后将最上部的信号强度、电信商、时间和电量信息都去掉，这部分信息由苹果系统自动在应用的最上部显示。

图 24.1　iOS 自带计算器界面

修改后的计算器页面素材如图 24.2 所示，将其保存好，后面将使用此图片作为背景图片。

有了界面后，按照第 23 章介绍的流程创建一个叫作 Calculator 的工程。

在文件导航中选中管理图片资源的文件夹 Assets.xcassets，可以看到里面还没有添加任何图片。打开刚才保存苹果计算器图片的位置，将该图片拖动到 Assets.xcassets 文件夹中，可以看到出现了 calculator 的图片文件，如果文件名不一致，请将其命名为 calculator。这样，我们就将苹果计算器图片的资源添加到工程中了，在工程中使用时只输入图片名即可，不需要再到图片的文件目录里去找。工程中用到的图片资源都可以通过这种方式加入工程中，由 Assets.xcassets 统一管理。例如，要添加一个应用的图标，可以将设计好的图片文件直接拖动到 AppIcon 中，如图 24.3 所示。

选中左侧导航栏中的文件 Main.storyboard，打开界面编辑器。在界面编辑器已

图 24.2　修改后的计算器页面素材

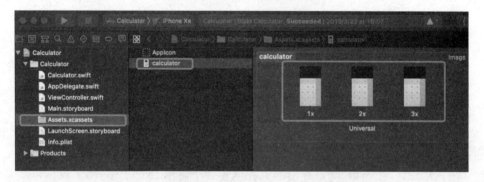

图 24.3　添加文件到图片资源文件夹中

经自动创建了视图控制器 ViewController，以及该控制器负责的视图，默认已经创建了一个 View。选中这个 View，通过拖动的方式，将组件库中的 UIImage View 组件拖到 View 中，如图 24.4 所示。

　　选中新添加的 UIImage View，在最右侧的属性设置器中设置属性 Image 的值为 calculator 即可看到刚刚放到 Assets.xcassets 中的苹果计算器图片已经出现了。细心的读者在动手操作的时候会发现，刚刚输入了 calculator 的前几个字母，系统就提供了完整的单词供选择，其实只要将图片添加到 Assets.xcassets 中，在使用该图片的

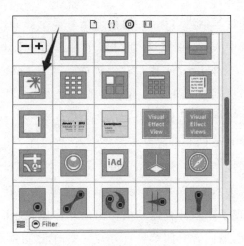

图 24.4　挑选 UIImageView

时候，系统就会启动联想功能，在用户只提供前几个字母的时候，将整个文件名显示出来供选择。因此，不仅要将应用中用到的图片都纳入 Assets.xcassets 中进行管理，还要给每一个图片文件起一个容易记忆和理解的名字，方便使用时直接调用。另外，还需要调整 Image View 的大小，通过鼠标拖动 4 个边角，调整其尺寸。这里将其大小调整到铺满全屏，如图 24.5 所示。

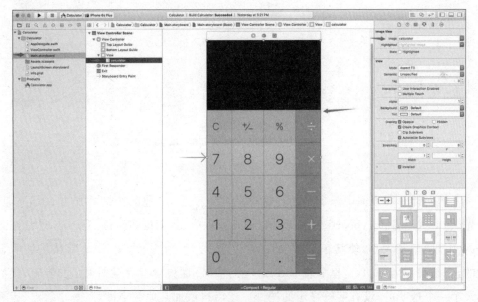

图 24.5　调整页面图片的大小

前面对截屏的苹果计算器进行编辑，去掉了结果显示部分，现在为应用添加这部分。如图 24.6 所示，选中 View 视图，在组件库中找到标签 Label，将其拖到 View 中的合适位置。调整 Label 的尺寸，使其左右都顶格。选中 Label 组件，打开属性编辑窗口，进行一系列设置，达到苹果计算器的结果显示的样式。首先，将 Text 属性的内容修改为 0，即计算输出结果的默认值为 0。字体颜色为黑色，要改为白色，将 Color 属性中的颜色修改为 White Color，这里不需要输入，直接在下拉列表中选择即可。字体太粗太小，将 Font 属性调整为 System Thin 100.0，其中 Thin 表示细体，100.0 表示字号。调整对齐格式为右对齐，将 Alignment 属性调整为第三个按钮处于选中状态。通过以上设置，实现了苹果计算器结果输出的字体样式。Label 的属性编辑器中还有很多其他高级属性，请读者参照苹果官方开发文档，这里不再一一介绍。

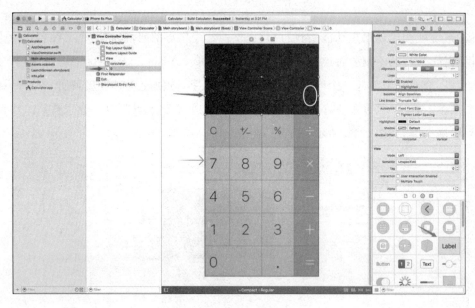

图 24.6　添加 Label 组件

下面向计算器界面中添加按键。这里只演示一个，其他的按键添加方式都类似。首先，选中 Main.storyboard 中的 View，打开组件库，找到按钮组件 Button，将其拖到 View 中，调整其大小，使其正好覆盖按键"4"如图 24.7 所示。Button 默认是透明显示的，正是我们需要的，这里只将 Button 中的 Title 属性设置为空即可。实际上，我们设置了一个透明的、没有字符显示的按键。然后再添加其他数字键，以及小数点、运算符、归零键等。到此为止，界面就设计完毕了。

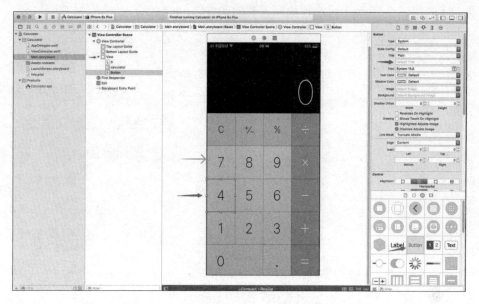

图 24.7 添加 Button

24.2 动作处理

在苹果系统中，每个用户的动作都会被系统所捕获。针对每个动作都有相应的事件处理。例如，用户按了计算机中的某个数字键，那么系统就需要保存这个数字键代表的值，这就需要在事件处理中完成。

下面为所有的按键添加动作事件处理。在 Xcode 中为一个动作增加事件处理是非常方便的事情，所有框架性的文件已经由 Xcode 自动生成了，我们需要做的就是建立动作与事件处理函数之间的关系。

如图 24.8 所示，要为界面中的具体组件建立动作处理，需要同时打开界面编辑器和代码编辑器。这里，首先在文件导航栏中选中 Main.storyboard，中间会显示界面编辑器。然后单击上部工具栏的右部，有一个两个圆形的按钮，单击该按钮，此时会在界面编辑器的右侧打开代码编辑器。这里默认打开的正是我们需要打开的 ViewController.swift。我们的工程比较简单，包含的文件比较少，如果遇到文件比较多的情况，很可能默认打开的代码文件不是我们需要的，不过没关系，可以单击代码编辑器上部的按钮 Automatic，选择我们需要的文件。

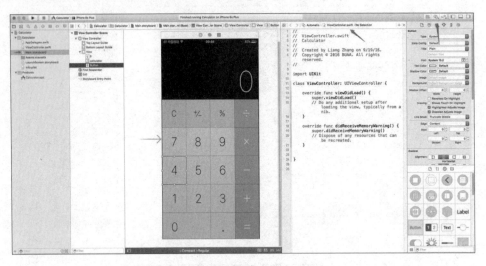

图 24.8　同时打开界面编辑器和代码编辑器

在图形组件浏览器中选中要添加动作处理的组件 Button，按住鼠标右键拖动到代码编辑中，然后松开鼠标右键，此时会弹出一个编辑框。这个编辑框既可以为图形组件指定一个动作处理，又可以将该组件作为一个输出变量增加到控制器类中。这里，从 Connection 的下拉列表中选择 Action，即动作处理。Name 为动作处理函数的名字，输入 btnPressed，最后单击 Connect，如图 24.9 所示。

如图 24.10 所示，系统自动为我们在 ViewController 类里面增加一个动作处理函数 btnPressed()。这里需要注意的是，动作处理函数和其他函数看上去有两点不同之处：第一个不同是在函数关键字前面有一个"@IBAction"，表示这是一个动作处理函数；第二个不同是在行号的前面有一个带实心点的圆圈，表示该动作处理函数已经和一个界面中的一个组件建立了联系，如果是一个空心圈，则表示还没有组件与之关联。

下面为其他的按键都建立动作处理。由于每个按键的处理过程都类似，所以不需要为每个按键都建立不同的动作处理函数。所有按键的动作都可以用函数 btnPressed() 处理。选中一个 button，按鼠标右键将其拖动到 btnPressed 的定义部分，当出现一个矩形框的时候释放鼠标右键。重复这个动作，为每个 Button 按键建立

第 24 章 计算器

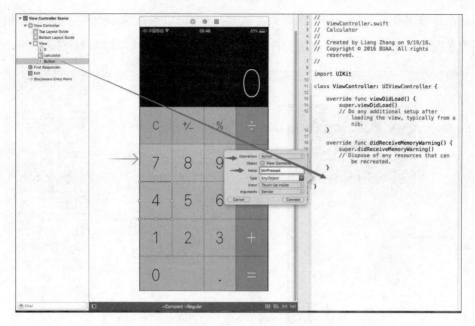

图 24.9 添加动作处理函数 btnPressed

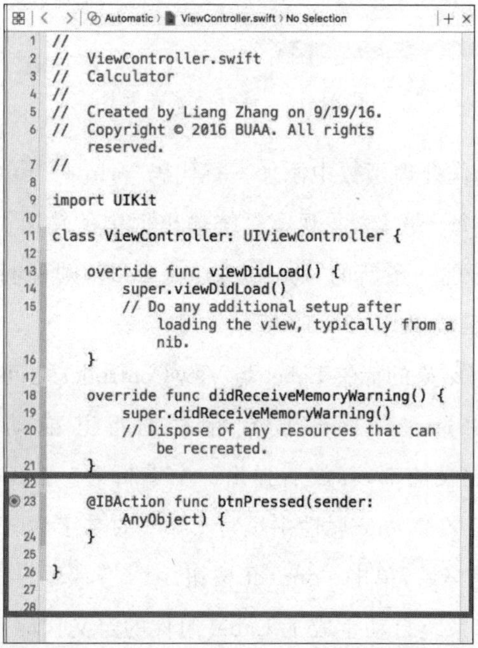

图 24.10 自动生成的动作处理代码

和动作处理函数 btnPressed() 的联系。连接完后,可以通过选中组件,查看右部的连接关系查看器确认是否正确地为组件建立了动作处理函数,如图 24.11 所示。

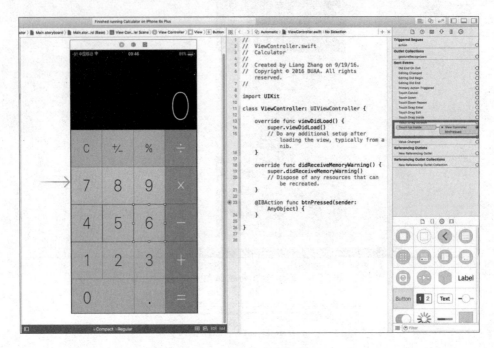

图 24.11　连接关系查看器

另外,还可以在动作处理函数中添加一行代码"print("button is pressed")"。然后启动应用,依次单击每个按键,并观察系统输出框中的输出。如果每个键被按的时候,系统输出框都会出现一条新的"button is pressed",则说明每个 button 组件已经正确建立了动作处理函数,如图 24.12 所示。

下面建立显示计算结果的标签 Label 与 ViewController 之间的联系。如图 24.13 所示,在界面浏览器中选择 Label,按鼠标右键将其拖动到代码编辑窗口的 ViewController 类的定义前部,释放后弹出一个编辑框。在编辑框中,Connection 属性选择默认的 Outlet。在 Name 属性中填写 Label 的名字为 resultLabel。Storage 属性选择 Weak。设置完毕后,单击 Connect 按钮。

如图 24.14 所示,系统自动生成了 Label 组件的定义代码。在类定义中,Label 是作为一个特殊的变量存在的。这个变量是 UILabel 类型的,当应用运行的时候,它和

第 24 章 计算器 Chapter 24

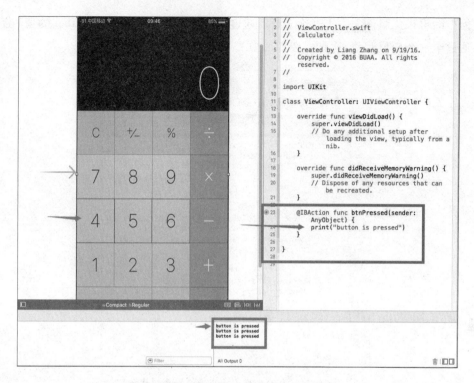

图 24.12 添加代码到动作处理函数中

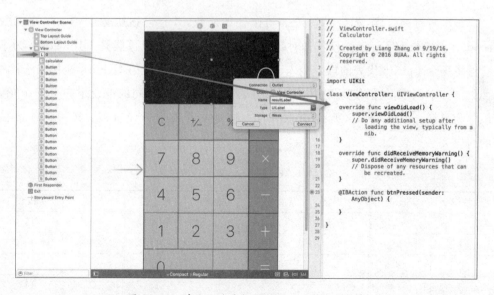

图 24.13 建立 Label 与 ViewController 的联系

界面中的可视化组件 Label 等价。可以通过这个变量读取到界面中 Label 的值，也可以通过给这个变量赋值，达到修改界面中 Label 值的目的。

```
import UIKit

class ViewController: UIViewController {

    @IBOutlet weak var resultLabel: UILabel!

    override func viewDidLoad() {
        super.viewDidLoad()
        // Do any additional setup after
            loading the view, typically from a
            nib.
    }

    override func didReceiveMemoryWarning() {
        super.didReceiveMemoryWarning()
        // Dispose of any resources that can
            be recreated.
    }

    @IBAction func btnPressed(sender:
        AnyObject) {

    }
}
```

图 24.14　自动生成 Label 组件的声明代码

每次按键的时候，都会触发按键动作处理函数，并且向该函数中传入一个参数 sender，该参数为实际触发动作的对象。在本例中，该对象只可能是 UIButton。为了能够将每个 Button 代表的内容传入动作处理函数中，必须利用 UIButton 中的一个属性记录。本例中，我们使用 UIButton 的属性 Tag，为每个不同的键设定一个 Tag 值，该值为整型。如图 24.15 所示，依次为每个按键的 Tag 属性设置值。按键与 Tag 值的对应关系见表 24.1。

表 24.1　按键与 Tag 值的对应关系

按键	Tag 值	按键	Tag 值	按键	Tag 值	按键	Tag 值
0	0	5	5	.	10	/	15
1	1	6	6	=	11	%	16
2	2	7	7	+	12	+/−	17
3	3	8	8	−	13	C	18
4	4	9	9	*	14		

第 24 章 计算器 Chapter 24

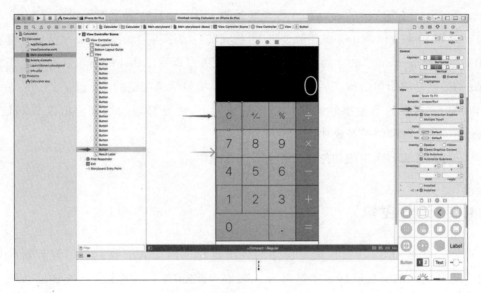

图 24.15 设置 Tag 属性

为每个按键添加完 Tag 属性的值后，要在动作处理函数 btnPressed() 中测试是否达到预期。如图 24.16 所示，向该函数中添加代码如下：

```swift
import UIKit

class ViewController: UIViewController {

    @IBOutlet weak var resultLabel: UILabel!

    override func viewDidLoad() {
        super.viewDidLoad()
        // Do any additional setup after loading the view, typically from a nib.
    }

    override func didReceiveMemoryWarning() {
        super.didReceiveMemoryWarning()
        // Dispose of any resources that can be recreated.
    }

    @IBAction func btnPressed(sender: AnyObject) {
        let btn: UIButton = sender as! UIButton
        resultLabel.text = String(btn.tag)
        print(resultLabel.text!)
    }

}
```

图 24.16 向函数 btnPressed() 中添加 Tag 属性的相关代码

第 26 行将传入的参数 sender 强制转换为 UIButton 类型,并赋值给变量 btn。

第 27 行将 btn 的 tag 属性值读出,再将其转换为字符串类型,然后赋值给计算输出结果的标签 resultLabel 的 text 属性,该属性负责保存 Label 的显示字符串。正常情况下,启动应用单击按键后会在计算输出结果中显示该按键的 tag 值。

第 28 行在计算输出结果显示不了的情况下,通过打印向系统输出该 Label 的 text 值,以检查赋值是否成功。

至此,计算器应用的动作处理相关的工作完毕。

24.3　运算逻辑

本节将运用 Swift 语言编写应用的处理逻辑。回忆一下我们使用计算器的场景:打开应用的时候默认显示结果为 0,现在已经做到了。单击数字按键输入第一个操作数,然后单击运算操作符,再接着输入第二个操作数,最后单击"等于"显示计算结果。

在这个过程中,所有类型的输入都是通过按键这个动作将 Button 的 tag 值传入处理函数的。那么,显示结果仅取决于即时的按键的 tag 值吗?答案是否定。显示的结果(不一定是计算结果,有可能还没有输入运算符)不仅取决于当前的输入,即时的按键的 tag 值,而且还取决于当前 Label 的 text 值。由这两个值决定显示的结果。因此,如果有一个负责运算的函数,就需要向它传入两个值:按键的 tag 值和当前 Label 的 text 值,该函数的返回值应该是计算后得到的结果。该结果将通过计算结果显示的 Label 输出到屏幕上。

单独建立一个 Swift 文件编写运算处理程序。选中文件浏览器中的项目名 Calculator,如图 24.17 所示,依次单击菜单栏中的 File→New→File,打开创建文件窗口。

如图 24.18 所示,在左侧的列表中选择 iOS 下的 Source,再在右边的列表中选择 Swift File,最后单击 Next 按钮。

如图 24.19 所示,输入文件名 Calculator,单击 Create 按钮。

如图 24.20 所示,系统自动生成文件 Calculator.swift,文件中只有一行代码,即导入 Foundation 框架文件。

第 24 章 计算器　Chapter 24

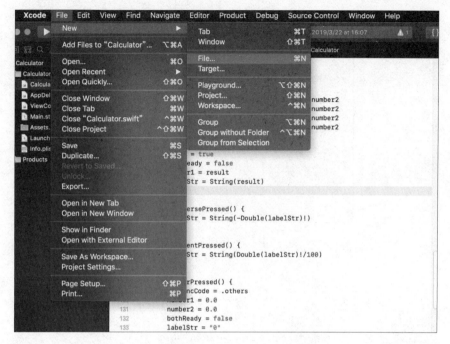

图 24.17　创建 Swift 文件

图 24.18　Swift File 选择页面

267

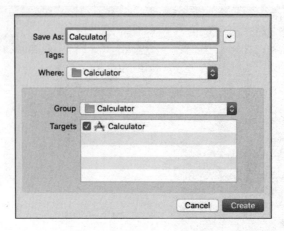

图 24.19　文件信息设置窗口

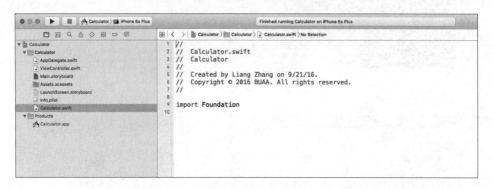

图 24.20　自动生成的文件 Calculator.swift

在文件 Calculator.swift 中创建一个 Calculator 类，用来处理相关的操作。类中含有一个方法，如图 24.21 所示。该方法有两个输入参数 tagValue 和 curResult，分别用来存储按键的 tag 属性值和当前计算输出结果 Label 的 text 属性值。返回值为字符型，即经过函数运算后得到的计算结果。目前该函数还只是一个框架，没有实质性的逻辑代码。我们添加两行代码用来测试：第 17 行打印传入的两个参数值；第 18 行将 curResult 作为返回值。

定义好 Calculator 类之后，就可以在 ViewController 中使用了。向文件 ViewController.swift 中添加如下代码（见图 24.22）：

第 13 行，声明一个 Calculator 类的变量 calculator。

```
 9  import Foundation
10
11  class Calculator {
12
13
14
15      func calculating(tagValue : Int, curResult : String) -> String {
16
17          print("tagValue is \(tagValue) and curResult is \(curResult)")
18          return curResult
19      }
20
21
22  }
```

图 24.21　定义函数 calculating()

```
 1  //
 2  //  ViewController.swift
 3  //  Calculator
 4  //
 5  //  Created by Liang Zhang on 9/19/16.
 6  //  Copyright © 2016 BUAA. All rights reserved.
 7  //
 8
 9  import UIKit
10
11  class ViewController: UIViewController {
12
13      var calculator : Calculator!    ←
14
15      @IBOutlet weak var resultLabel: UILabel!
16
17      override func viewDidLoad() {
18          super.viewDidLoad()
19          // Do any additional setup after loading the view, typically from a nib.
20          calculator = Calculator()    ←
21
22      }
23
24      override func didReceiveMemoryWarning() {
25          super.didReceiveMemoryWarning()
26          // Dispose of any resources that can be recreated.
27          calculator = nil    ←
28      }
29
30      @IBAction func btnPressed(sender: AnyObject) {
31          let btn: UIButton = sender as! UIButton
32          resultLabel.text = calculator.calculating(btn.tag, curResult: resultLabel.text!)  ←
33      }
34
35  }
36
```

图 24.22　向 ViewController 中添加代码

第 20 行，在视图加载的时候创建一个类 Calculator 的实例，并将其赋值给变量 calculator。需要说明一下，viewDidLoad() 方法相当于 ViewController 的初始化方法，所以类实例的创建需要在该方法中完成。

第 27 行，在视图销毁前，将实例 calculator 销毁掉，以释放其所占的内存。

第 32 行，调用实例 calculator 中的方法 calculating()，并将按键的 tag 值和计算结果输出 Label 的 text 值作为参数传入,返回的计算结果重新再赋值给计算结果输出 Label 的 text 属性。

单击"运行"按钮,在模拟器中单击按键"1"和"2",观察系统输出窗口,如图 24.23 所示,结果正确,说明前面的代码达到预期效果。下面再进一步细化。

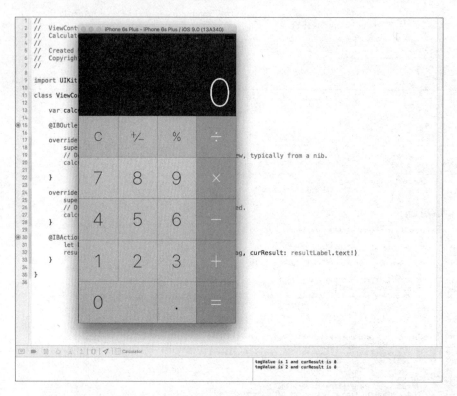

图 24.23 运行结果

如图 24.24 所示,在类 Calculator 中定义好成员变量,并在构造器中对成员变量进行初始化操作。这里列出了析构器,但本例并没有需要析构器释放的资源,故只向系统打印显示一个调用析构器的信息。

类中一共定义了 7 个成员变量,通过这 7 个成员变量可以将计算器的重要数据信息记录下来,在类中的方法中可以对它们进行修改。成员变量的含义和作用如下所示。

curFuncCode:该变量为 FuncCode 枚举型,用来记录当前有效的双目运算符,如果当前没有有效的双目运算符,curFuncCode 的取值为默认的".others"。

```
19  class Calculator {
20
21      var curFuncCode : FuncCode
22      var number1 : Double
23      var number2 : Double
24      var labelStr : String
25      var bothReady : Bool
26      var tagValue : Int
27      var start : Bool
28
29      init(){
30          curFuncCode = .others
31          number1 = 0.0
32          number2 = 0.0
33          bothReady = false
34          labelStr = "0"
35          tagValue = 0
36          start = true
37          print("app is inited")
38      }
39
40      deinit{
41          print("app is deinited")
42      }
```

图 24.24 定义类 Calculator

number1：该变量为双精度浮点型，用来保存双目运算的第一个操作数。

number2：该变量为双精度浮点型，用来保存双目运算的第二个操作数。

labelStr：该变量为字符串型，用来保存当前计算器中的显示计算结果的 Label 组件的 text 属性值。

bothReady：该变量为布尔型，用来标识当前是否为双目运算。

tagValue：该变量为整型，用来记录按键 UIButton 的 tag 属性值。

start：该变量为布尔型，用来标识是否开始输入数字。

所有键都是通过 button 的 tag 属性值传入的，那么就需要对 tag 值进行解析。对其解析的第一步是分类，大致可分为 7 类：数字"0~9"，等号"="，归零"C"，点号"."，两个操作数的四则运算"+、-、*、/"，符号取反"+/-"，取百分数"%"。其中数字比较简单，其 tag 值和数值相等。其他的都属于功能键。

其中的四则运算可以通过一个枚举型 FuncCode，将功能与 tag 值的对应关系描述出来，后面在解析 tag 值的时候可以直接使用枚举值进行匹配，大大增加了代码的可读性。图 24.25 所示为枚举型 FuncCode 的定义。

```
11  enum FuncCode : Int {
12      case plus = 12
13      case minus = 13
14      case multiply = 14
15      case divide = 15
16      case others = 100
17  }
```

图 24.25 枚举型 FuncCode 的定义

如图 24.25 所示，列出了双目操作的加减乘除和一个默认值 others。当没有双目操作的时候，取默认值。

每个不同类型的按键都需要进行不同的处理，为此定义了不同的处理函数，如图 24.26 所示，这里只有函数名，具体的函数体在后面完善。

如图 24.27 所示，完善 calculating()方法。首先，判断当前运算符是否为双目运算符，即是否为加、减、乘、除运算。如果不是双目运算符，则将当前的运算结果输出的 Label 的值 curLabel 赋值给类成员变量 labelStr。如果是双目运算符，则将赋值类成员变量 labelStr 的工作放到具体的处理方法中，这里暂不处理。接着，获取按键 Button 的 tag 值，将传入参数 tagValue 赋值给类成员变量 tagValue。

然后通过 switch 结果解析不同的按键类型，根据 tagValue 的值，分类进行处理，对于不同类型采取不同的处理函数。当 tagValue 为 0~9 时，显然按键为数字键，调用数字键处理方法 numberPressed()；当 tagValue 为 10 时，按键为点号键，调用点号键处理方法 dotPressed()；当 tagValue 为 11 时，按键为等于键，调用等于键处理方法 equalPressed()；当 tagValue 为 12~15 时，按键为双目运算键，调用双目运算键处理

```
func numberPressed(){
}
func dotPressed(){
}
func operatorPressed(){
}
func equalPressed(){
}
func conversePressed(){
}
func percentPressed(){
}
func clearPressed(){
}
```

图 24.26　定义函数接口

```
func calculating(tagValue : Int, curLabel : String) -> String {
    if curFuncCode == .others {
        labelStr = curLabel
    }
    self.tagValue = tagValue
    switch tagValue {
        case 0...9 : numberPressed()
        case 10 : dotPressed()
        case 11 : equalPressed()
        case 12...15 : operatorPressed()
        case 16 : percentPressed()
        case 17 : conversePressed()
        case 18 : clearPressed()
        default : return labelStr
    }
    if labelStr.hasSuffix(".0") {
        labelStr = labelStr.substringToIndex(labelStr.endIndex.predecessor().predecessor())
    }
    return labelStr
}
```

图 24.27　实现 calculating()函数

方法 operatorPressed()；当 tagValue 为 16 时，按键为取百分比键，调用取百分比键处理方法 percentPressed()；当 tagValue 为 17 时，按键为符号取反键，调用符号取反键处理方法 conversePressed()；当 tagValue 为 18 时，按键为归零键，调用归零键处理方法 clearPressed()；默认情况下，直接返回 labelStr。

在返回 labelStr 前，要对 labelStr 进行简单的处理。在类 Calculator 中，我们定义了成员变量 number1、number2 保存操作数，类型都是双精度浮点数，都含有小数点，所以运算结果也含有小数点。当结果的小数部分为 0 时，就应该去掉".0"，只显示整数部分。Swift 的 String 类型有一个很有用的方法 hasSuffix()，会判断字符串的后缀是否还有特定的子串，这里就是用来判断运算结果的字符串中是否含有后缀".0"，如果含有这个后缀，则可以去掉它，只保留整数部分。为了去掉运算结果字符串的最后两个字符，要用到 String 类型的属性 endIndex，先取出最后一个字符的索引值，然后通过方法 predecessor() 再向前倒退两位，即取倒数第 3 个字符的索引值。然后再将此索引值作为 String 类型中方法 substringToIndex() 的参数传入，获取从第一个字符开始至倒数第 3 个字符为止的子串，从而截取字符串的整数部分。

如图 24.28 所示，实现 numberPressed() 函数。当按键为 0~9 的数字时，会触发该函数。该函数的处理有以下两种情况。

```
70  func numberPressed(){
71
72      if ((labelStr == "0") || start ) {
73          labelStr = String(tagValue)
74          start = false
75      } else {
76          labelStr = labelStr + String(tagValue)
77      }
78
79  }
```

图 24.28　实现 numberPressed() 函数

（1）如果当前显示字符串为 0 或者要开始输入一个新数时，将按键值的字符串直接赋值给当前的显示字符串，然后再将输入新数的标志 start 置为 false。

（2）如果不满足上述条件，就将当前的显示字符串和按键值的字符串连接起来。

如图 24.29 所示，实现 clearPressed() 函数。该函数的功能非常简单，将构造器里的初始化语

```
134  func clearPressed(){
135      curFuncCode = .others
136      number1 = 0.0
137      number2 = 0.0
138      bothReady = false
139      labelStr = "0"
140      tagValue = 0
141      start = true
142  }
```

图 24.29　实现 clearPressed() 函数

句再执行一遍,特别是将显示运算结果的 labelStr 修改为"0"。

如图 24.30 所示,实现 percentPressed() 函数。该函数的功能也非常简单,即将当前的计算结果转换为浮点数,然后除以 100,再转换为字符串赋值给 labelStr。

```
130   func percentPressed(){
131       labelStr = String(Double(labelStr)!/100)
132   }
```

图 24.30　实现 percentPressed() 函数

如图 24.31 所示,实现 conversePressed() 函数。该函数的功能是为当前值的符号取反,现将 labelStr 转换为浮点型,再符号位取反,最后转换为字符串赋值给 labelStr。

```
126   func conversePressed(){
127       labelStr = String(-Double(labelStr)!)
128   }
```

图 24.31　实现 conversePressed() 函数

如图 24.32 所示,实现 dotPressed() 函数。该函数的功能是为当前值加一个小数点。这里需要注意一点,如果当前值已经包含小数点,则不做任何改变,返回原值;如果当前值不含小数点,则在当前值的字符串基础上附加一个小数点即可。

```
80    func dotPressed(){
81        if !labelStr.containsString(".") {
82            labelStr = labelStr + "."
83        }
84    }
```

图 24.32　实现 dotPressed() 函数

如图 24.33 所示,实现 operatorPressed() 函数。该函数的功能是处理按键为双目运算符的情况。当按键为加、减、乘、除运算符的时候,有两种情况:第一种情况是按键前输入了一个操作数;第二种情况是按键前输入两个操作数和一个加、减、乘、除运算符。这里用类成员变量 bothReady 区别这种情况。

当 bothReady 为 false 时,为第一种情况,做如下操作:

(1) 根据 tagValue 的值取到 FuncCode 类型的枚举值,并赋值给 curFuncCode。

(2) 将当前显示的计算结果 labelStr 转换为浮点型,并赋值给 number1 保存起来,以备运算时使用。

```
86      func operatorPressed() {
87          if bothReady {
88              var result : Double
89              result = 0.0
90              number2 = Double(labelStr)!
91              switch curFuncCode {
92              case .plus :       result = number1 + number2
93              case .minus :      result = number1 - number2
94              case .multiply :   result = number1 * number2
95              case .divide :     result = number1 / number2
96              case .others :     print("error")
97              }
98              start = true
99              number1 = result
100             labelStr = String(result)
101         } else {
102             curFuncCode = FuncCode(rawValue: tagValue)!
103             number1 = Double(labelStr)!
104             bothReady = true
105             start = true
106         }
107     }
```

图 24.33　实现 operatorPressed() 函数

（3）将 bothReady 赋值为 true，说明已经输入了一个运算符，如果再次出现新的运算符，就要进入计算环节了。

（4）将 start 标识赋值为 true，说明可以输入第 2 个数了。

当 bothReady 为 true 时，为第二种情况，做如下操作：

（1）声明一个局部变量 result，类型为双精度浮点型，用来记录运算结果。

（2）result 赋初值为 0.0。

（3）当前的 labelStr 转换为浮点型，赋值给 number2，即作为第 2 个操作数。

（4）通过 switch 语句匹配 curFuncCode，确定具体的运算符。匹配后，进行四则运算，并将结果保留到 result 中。

（5）设置 start 为 true，准备接收新输入的数字。

（6）将计算结果 result 保存到 number1 中，为下一次运算做好准备。

（7）将运算结果 result 转换为字符串，并赋值给 labelStr。

如图 24.34 所示，实现 equalPressed() 函数。该函数的功能为处理等号键被按下后的操作。在函数体中做如下操作：

（1）声明局部变量 result，类型为双精度浮点型，用来保存运算结果。

（2）变量 result 赋初始值为 0.0。

（3）当前的 labelStr 转换为浮点型，赋值给 number2，即作为第 2 个操作数。

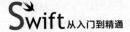

```
109    func equalPressed() {
110        var result : Double
111        result = 0.0
112        number2 = Double(labelStr)!
113        switch curFuncCode {
114            case .plus :       result = number1 + number2
115            case .minus :      result = number1 - number2
116            case .multiply :   result = number1 * number2
117            case .divide :     result = number1 / number2
118            case .others :     print("error")
119        }
120        start = true
121        bothReady = false
122        number1 = result
123        labelStr = String(result)
124    }
```

图 24.34 实现 equalPressed()函数

（4）通过 switch 语句匹配 curFuncCode，确定具体的运算符。匹配后，进行四则运算，并将结果保留到 result 中。

（5）设置 start 为 true，准备接收新输入的数字。

（6）将 bothReady 赋值为 false，说明前面运算已经结束，之前保存的 curFuncCode 不在下次运算中生效。

（7）将计算结果 result 保存到 number1 中，为下次运算做好准备。

（8）将运算结果 result 转换为字符串，并赋值给 labelStr。

24.4 小结

本章通过一个实际的应用工程，综合运用 Swift 语言和 Xcode 的 Interface Builder 实现了苹果系统自带计算器的功能。对于如何构造苹果应用的界面以及如何调用苹果提供的各种强大的 API，超出了本书的范畴。本章中尽量避免读者陷入 Swift 语言以外的内容中，我们主要介绍如何运用 Swift 语言解决实际问题，使读者在实际的例子中对 Swift 语言有一个更综合的认识。对 Swift 语言的学习是一个循序渐进的过程，只有把语言基础掌握牢固，才可能在进一步运用语言开发实际应用中游刃有余。

第 25 章
手游 2048

手游 2048 是一款流行的手机数字游戏，由 Gabriele Cirulli 开发，第一个版本发布于 2014 年 3 月 20 日。随着手游 2048 在 iOS 上的爆红，Android 版和网页版的游戏 2048 也随之出现。在原作的基础上，还出现了不少改进版的手游。

手游 2048 如图 25.1 所示，操作界面为一个正方形，由 16 个方形格子组成。每个格子可以放置一个标有数字的砖块。初始时，有两个格子中已经放入了两个数字为 2 的砖块。

手游 2048 的游戏规则：选择上、下、左、右滑动砖块。每滑动一次，所有的数字方块都会往滑动的方向靠拢，系统也会在空白的地方随机出现一个数字砖块（数字为 2 或者为 4）。相同的数字砖块在

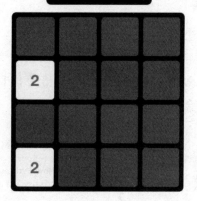

图 25.1　手游 2048

靠拢相撞时会合并相加为一个砖块。通过反复滑动砖块实现相同数字砖块的合并，最终得到 2048 的数字砖块时就算成功。

手游 2048 的设计采用 MVC 架构，即模型层—视图层—控制层。其中，模型层用来抽象出游戏中的数据实体和运行逻辑，视图层用来表示交互界面中的各个视图元素，控制层用来控制视图的显示以及视图层与模型层的交互。下面依次从模型层、视图层和控制层实现该应用。

25.1 模型层

模型层描述游戏的抽象模型及其内在逻辑。在项目中创建辅助模型文件 AuxiliaryModels.swift 和游戏模型文件 GameModel.swift。其中，辅助模型文件用来定义模型中相关的结构体和枚举类型，而游戏模型文件则用来抽象游戏模型和规则（即砖块的移动、合并等算法）。

25.1.1 辅助模型

如图 25.2 所示，在辅助模型文件中定义枚举类型 MoveDirection，用来描述砖块移动的 4 个方向，即向上、向下、向左和向右。

```
enum MoveDirection {
    case up, down, left, right
}
```

图 25.2 枚举类型 MoveDirection

如图 25.3 所示，在辅助模型文件中定义枚举类型 MoveOrder，用来描述两种移动指令（移动和合并）的相关信息，包括砖块的原位置、目标位置、当前值和是否合并。

```
enum MoveOrder {
    case singleMoveOrder(source: Int, destination: Int, value: Int,
        wasMerge: Bool)
    case doubleMoveOrder(firstSource: Int, secondSource: Int,
        destination: Int, value: Int)
}
```

图 25.3 枚举类型 MoveOrder

如图 25.4 所示，在辅助模型文件中定义枚举类型 TileObject，用来描述游戏板中砖块位置的两种情况，即无砖块和有砖块（含当前砖块值）。

```
enum TileObject {
    case empty
    case tile(Int)
}
```

图 25.4 枚举类型 TileObject

如图 25.5 所示，在辅助模型文件中定义枚举类型 ActionToken，用来描述游戏中的手势标识，包括无动作、移动和两种合并动作。

```swift
enum ActionToken {
  case noAction(source: Int, value: Int)
  case move(source: Int, value: Int)
  case singleCombine(source: Int, value: Int)
  case doubleCombine(source: Int, second: Int, value: Int)

  // Get the 'value', regardless of the specific type
  func getValue() -> Int {
    switch self {
    case let .noAction(_, v): return v
    case let .move(_, v): return v
    case let .singleCombine(_, v): return v
    case let .doubleCombine(_, _, v): return v
    }
  }
  // Get the 'source', regardless of the specific type
  func getSource() -> Int {
    switch self {
    case let .noAction(s, _): return s
    case let .move(s, _): return s
    case let .singleCombine(s, _): return s
    case let .doubleCombine(s, _, _): return s
    }
  }
}
```

图 25.5　枚举类型 ActionToken

如图 25.6 所示，在辅助模型文件中定义结构体 MoveCommand，用来描述移动砖块命令，包括移动方向和移动结果。

```swift
struct MoveCommand {
  let direction : MoveDirection
  let completion : (Bool) -> ()
}
```

图 25.6　结构体 MoveCommand

如图 25.7 所示，在辅助模型文件中定义结构体 SquareGameboard，用来描述游戏板，包括：游戏板的尺寸和用来存储每个格子的数组。该结构体为泛型，每个格子的内容可以为任意类型。在本应用中，每个格子的内容为 Int 类型。

25.1.2　游戏模型

如图 25.8 所示，在游戏模型文件中定义协议 GameModelProtocol，用来定义模型层和控制层之间通信的接口函数，包括砖块值改变函数 scoreChanged()、移动一块砖

```
struct SquareGameboard<T> {
  let dimension : Int
  var boardArray : [T]

  init(dimension d: Int, initialValue: T) {
    dimension = d
    boardArray = [T](repeating: initialValue, count: d*d)
  }

  subscript(row: Int, col: Int) -> T {
    get {
      assert(row >= 0 && row < dimension)
      assert(col >= 0 && col < dimension)
      return boardArray[row*dimension + col]
    }
    set {
      assert(row >= 0 && row < dimension)
      assert(col >= 0 && col < dimension)
      boardArray[row*dimension + col] = newValue
    }
  }

  mutating func setAll(to item: T) {
    for i in 0..<dimension {
      for j in 0..<dimension {
        self[i, j] = item
      }
    }
  }
}
```

图 25.7　结构体 SquareGameboard

的函数 moveOneTile()、移动两块砖的函数 moveTwoTiles() 和插入砖块函数 insertTile()。

```
protocol GameModelProtocol : class {
  func scoreChanged(to score: Int)
  func moveOneTile(from: (Int, Int), to: (Int, Int), value: Int)
  func moveTwoTiles(from: ((Int, Int), (Int, Int)), to: (Int, Int),
      value: Int)
  func insertTile(at location: (Int, Int), withValue value: Int)
}
```

图 25.8　协议 GameModelProtocol

　　在游戏模型文件中定义类 GameModel，用来描述游戏的当前状态和运行逻辑。如图 25.9 所示，类 GameModel 中定义了游戏板的尺寸、阈值、当前分数等，并提供了 GameModel 中各个属性的初始化方法 init()。

```swift
class GameModel : NSObject {
  let dimension : Int
  let threshold : Int

  var score : Int = 0 {
    didSet {
      delegate.scoreChanged(to: score)
    }
  }
  var gameboard: SquareGameboard<TileObject>

  unowned let delegate : GameModelProtocol

  var queue: [MoveCommand]
  var timer: Timer

  let maxCommands = 100
  let queueDelay = 0.3

  init(dimension d: Int, threshold t: Int, delegate:
     GameModelProtocol) {
    dimension = d
    threshold = t
    self.delegate = delegate
    queue = [MoveCommand]()
    timer = Timer()
    gameboard = SquareGameboard(dimension: d, initialValue: .empty)
    super.init()
  }
```

图 25.9　类 GameModel

如图 25.10 所示，在类 GameModel 中定义函数 reset()，用来重置游戏板的状态，包括清零当前分数、清空游戏板中的砖块、清除队列和计时器失效。

```swift
func reset() {
  score = 0
  gameboard.setAll(to: .empty)
  queue.removeAll(keepingCapacity: true)
  timer.invalidate()
}
```

图 25.10　函数 reset()

如图 25.11 所示，在类 GameModel 中定义函数 queueMove()，用来实现砖块的移动操作，以响应用户的滑动手势。

```swift
func queueMove(direction: MoveDirection, onCompletion: @escaping
    (Bool) -> ()) {
  guard queue.count <= maxCommands else {
    return
  }
  queue.append(MoveCommand(direction: direction, completion:
     onCompletion))
  if !timer.isValid {
    timerFired(timer)
  }
}
```

图 25.11　函数 queueMove()

如图 25.12 所示，在类 GameModel 中定义函数 timerFired()，用来启动计时器。一旦计时器启动，游戏模型将执行移动操作并改变游戏的当前状态。

```swift
@objc func timerFired(_: Timer) {
  if queue.count == 0 {
    return
  }
  var changed = false
  while queue.count > 0 {
    let command = queue[0]
    queue.remove(at: 0)
    changed = performMove(direction: command.direction)
    command.completion(changed)
    if changed {
      break
    }
  }
  if changed {
    timer = Timer.scheduledTimer(timeInterval: queueDelay,
      target: self,
      selector:
      #selector(GameModel.timerFired(_:)),
      userInfo: nil,
      repeats: false)
  }
}
```

图 25.12 函数 timerFired()

如图 25.13 所示，在类 GameModel 中定义函数 insertTile()，用来实现在游戏板上的特定位置插入一个特定值的砖块。

```swift
func insertTile(at location: (Int, Int), value: Int) {
  let (x, y) = location
  if case .empty = gameboard[x, y] {
    gameboard[x, y] = TileObject.tile(value)
    delegate.insertTile(at: location, withValue: value)
  }
}
```

图 25.13 函数 inertTile()

如图 25.14 所示，在类 GameModel 中定义函数 insertTileAtRandomLocation()，用来实现在游戏板上的随机位置插入一个特定值的砖块。

```swift
func insertTileAtRandomLocation(withValue value: Int) {
  let openSpots = gameboardEmptySpots()
  if openSpots.isEmpty {
    return
  }
  let idx = Int(arc4random_uniform(UInt32(openSpots.count-1)))
  let (x, y) = openSpots[idx]
  insertTile(at: (x, y), value: value)
}
```

图 25.14 函数 insertTileAtRandomLocation()

如图 25.15 所示，在类 GameModel 中定义函数 gameboardEmptySpots()，通过元组型数组返回当前游戏板上所有空格子的坐标位置。

```swift
func gameboardEmptySpots() -> [(Int, Int)] {
  var buffer : [(Int, Int)] = []
  for i in 0..<dimension {
    for j in 0..<dimension {
      if case .empty = gameboard[i, j] {
        buffer += [(i, j)]
      }
    }
  }
  return buffer
}
```

图 25.15　函数 gameboardEmptySpots()

如图 25.16 所示，在类 GameModel 中定义函数 tileBelowHasSameValue()，用来判断砖块下方是否含有相同值的砖块。

```swift
func tileBelowHasSameValue(location: (Int, Int), value: Int) ->
    Bool {
  let (x, y) = location
  guard y != dimension - 1 else {
    return false
  }
  if case let .tile(v) = gameboard[x, y+1] {
    return v == value
  }
  return false
}
```

图 25.16　函数 tileBelowHasSameValue()

如图 25.17 所示，在类 GameModel 中定义函数 tileToRightHasSameValue()，用来判断砖块右边是否有相同值的砖块。

```swift
func tileToRightHasSameValue(location: (Int, Int), value: Int) ->
    Bool {
  let (x, y) = location
  guard x != dimension - 1 else {
    return false
  }
  if case let .tile(v) = gameboard[x+1, y] {
    return v == value
  }
  return false
}
```

图 25.17　函数 tileToRightHasSameValue()

如图 25.18 所示，在类 GameModel 中定义函数 userHasLost()，用来判断用户当前的游戏状态是否为失败，即全部空格已经填满砖块，并且无法继续合并。

```swift
func userHasLost() -> Bool {
  guard gameboardEmptySpots().isEmpty else {
    return false
  }

  for i in 0..<dimension {
    for j in 0..<dimension {
      switch gameboard[i, j] {
      case .empty:
        assert(false, "Gameboard reported itself as full, but we
            still found an empty tile. This is a logic error.")
      case let .tile(v):
        if tileBelowHasSameValue(location: (i, j), value: v) ||
          tileToRightHasSameValue(location: (i, j), value: v)
        {
          return false
        }
      }
    }
  }
  return true
}
```

图 25.18　函数 userHasLost()

如图 25.19 所示，在类 GameModel 中定义函数 userHasWon()，用来判断用户的当前游戏状态是否为获胜，即已经达到了游戏的阈值 2048。

```swift
func userHasWon() -> (Bool, (Int, Int)?) {
  for i in 0..<dimension {
    for j in 0..<dimension {
      if case let .tile(v) = gameboard[i, j], v >= threshold {
        return (true, (i, j))
      }
    }
  }
  return (false, nil)
}
```

图 25.19　函数 userHasWon()

如图 25.20 所示，在类 GameModel 中定义函数 performMove()，用来实现砖块移动的相关操作，包括移动后所有砖块数值的计算、游戏状态的更新等。

如图 25.21 所示，在类 GameModel 中定义函数 condense()，用来实现在砖块移动时，压缩两个砖块之间空格的操作。

如图 25.22 所示，在类 GameModel 中定义函数 quiescentTileStillQuiescent()，用

来判断一个没有移动的砖块是否仍然保持原来的位置。

```swift
func performMove(direction: MoveDirection) -> Bool {
  let coordinateGenerator: (Int) -> [(Int, Int)] = { (iteration:
      Int) -> [(Int, Int)] in
    var buffer = Array<(Int, Int)>(repeating: (0, 0), count:
        self.dimension)
    for i in 0..<self.dimension {
      switch direction {
      case .up: buffer[i] = (i, iteration)
      case .down: buffer[i] = (self.dimension - i - 1, iteration)
      case .left: buffer[i] = (iteration, i)
      case .right: buffer[i] = (iteration, self.dimension - i - 1)
      }
    }
    return buffer
  }

  var atLeastOneMove = false
  for i in 0..<dimension {
    let coords = coordinateGenerator(i)
    let tiles = coords.map() { (c: (Int, Int)) -> TileObject in
      let (x, y) = c
      return self.gameboard[x, y]
    }
    let orders = merge(tiles)
    atLeastOneMove = orders.count > 0 ? true : atLeastOneMove
    for object in orders {
      switch object {
      case let MoveOrder.singleMoveOrder(s, d, v, wasMerge):
        let (sx, sy) = coords[s]
        let (dx, dy) = coords[d]
        if wasMerge {
          score += v
        }
        gameboard[sx, sy] = TileObject.empty
        gameboard[dx, dy] = TileObject.tile(v)
        delegate.moveOneTile(from: coords[s], to: coords[d],
            value: v)
      case let MoveOrder.doubleMoveOrder(s1, s2, d, v):
        let (s1x, s1y) = coords[s1]
        let (s2x, s2y) = coords[s2]
        let (dx, dy) = coords[d]
        score += v
        gameboard[s1x, s1y] = TileObject.empty
        gameboard[s2x, s2y] = TileObject.empty
        gameboard[dx, dy] = TileObject.tile(v)
        delegate.moveTwoTiles(from: (coords[s1], coords[s2]), to:
            coords[d], value: v)
      }
    }
  }
  return atLeastOneMove
}
```

图 25.20　函数 performMove()

```
func condense(_ group: [TileObject]) -> [ActionToken] {
  var tokenBuffer = [ActionToken]()
  for (idx, tile) in group.enumerated() {
    switch tile {
    case let .tile(value) where tokenBuffer.count == idx:
      tokenBuffer.append(ActionToken.noAction(source: idx, value:
          value))
    case let .tile(value):
      tokenBuffer.append(ActionToken.move(source: idx, value:
          value))
    default:
      break
    }
  }
  return tokenBuffer;
}
```

图 25.21 函数 condense()

```
class func quiescentTileStillQuiescent(inputPosition: Int,
    outputLength: Int, originalPosition: Int) -> Bool {
  return (inputPosition == outputLength) && (originalPosition ==
      inputPosition)
}
```

图 25.22 函数 quiescentTileStillQuiescent()

如图 25.23 所示,在类 GameModel 中定义函数 collapse(),用来计算将一排砖块进行合并后的结果,包括合并的砖块及其值。

```
func collapse(_ group: [ActionToken]) -> [ActionToken] {
  var tokenBuffer = [ActionToken]()
  var skipNext = false
  for (idx, token) in group.enumerated() {
    if skipNext {
      skipNext = false
      continue
    }
    switch token {
    case .singleCombine:
      assert(false, "Cannot have single combine token in input")
    case .doubleCombine:
      assert(false, "Cannot have double combine token in input")
    case let .noAction(s, v)
      where (idx < group.count-1
        && v == group[idx+1].getValue()
        && GameModel.quiescentTileStillQuiescent(inputPosition:
            idx, outputLength: tokenBuffer.count,
            originalPosition: s)):
      let next = group[idx+1]
      let nv = v + group[idx+1].getValue()
```

图 25.23 函数 collapse()

```
          skipNext = true
          tokenBuffer.append(ActionToken.singleCombine(source:
              next.getSource(), value: nv))
        case let t where (idx < group.count-1 && t.getValue() ==
            group[idx+1].getValue()):
          let next = group[idx+1]
          let nv = t.getValue() + group[idx+1].getValue()
          skipNext = true
          tokenBuffer.append(ActionToken.doubleCombine(source:
              t.getSource(), second: next.getSource(), value: nv))
        case let .noAction(s, v) where
            !GameModel.quiescentTileStillQuiescent(inputPosition: idx,
            outputLength: tokenBuffer.count, originalPosition: s):
          tokenBuffer.append(ActionToken.move(source: s, value: v))
        case let .noAction(s, v):
          tokenBuffer.append(ActionToken.noAction(source: s, value: v))
        case let .move(s, v):
          tokenBuffer.append(ActionToken.move(source: s, value: v))
        default:
          break
        }
      }
      return tokenBuffer
    }
```

图 25.23 （续）

如图 25.24 所示，在类 GameModel 中定义函数 convert()。当计算一排砖块合并的移动结果时，该函数将 condense() 获取的 ActionToken 数组转化为 MoveOrder 数组，并将其返回给代理对象。

```
    func convert(_ group: [ActionToken]) -> [MoveOrder] {
      var moveBuffer = [MoveOrder]()
      for (idx, t) in group.enumerated() {
        switch t {
        case let .move(s, v):
          moveBuffer.append(MoveOrder.singleMoveOrder(source: s,
              destination: idx, value: v, wasMerge: false))
        case let .singleCombine(s, v):
          moveBuffer.append(MoveOrder.singleMoveOrder(source: s,
              destination: idx, value: v, wasMerge: true))
        case let .doubleCombine(s1, s2, v):
          moveBuffer.append(MoveOrder.doubleMoveOrder(firstSource: s1,
              secondSource: s2, destination: idx, value: v))
        default:
          break
        }
      }
      return moveBuffer
    }
```

图 25.24 函数 convert()

如图 25.25 所示,在类 GameModel 中定义函数 merge(),用来将一个给定的砖块对象数组进行合并,并返回一个对应的移动指令数组。

```
func merge(_ group: [TileObject]) -> [MoveOrder] {
    return convert(collapse(condense(group)))
}
```

图 25.25 函数 merge()

25.2 视图层

手游 2048 的操作界面比较简单。首先,通过 Launch Screen.storyboard 文件设置启动页面,如图 25.26 所示。启动页面是由项目自动创建的,只向其中添加控件即可。这里只向启动页面中添加一个 Label,用来显示游戏名 swift-2048。

图 25.26 启动页面

主故事板 Main.storyboard 由项目自动创建，在本应用中用来启动游戏。如图 25.27 所示，主故事板中只有一个按钮 Start Game，用来触发游戏启动。

图 25.27 故事板

游戏界面主要由 3 部分组成，分别为砖块视图、辅助视图（计分板）、游戏板视图。通常单独为每个视图编写视图文件。

25.2.1 砖块视图

创建砖块视图文件 TileView.swift，如图 25.28 所示。该文件中定义了一个类 TileView，用来描述砖块视图。该类包括砖块的数值属性、数字标签以及初始化方法。

25.2.2 辅助视图

创建辅助视图文件 AccessoryView.swift。该文件由 3 部分组成，分别为计分板视图协议 ScoreViewProtocol、计分板视图 ScoreView 以及控制视图 ControlView。

如图 25.29 所示，计分板视图协议提供了改变计分板分值的函数接口。

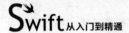

```
class TileView : UIView {
  var value : Int = 0 {
    didSet {
      backgroundColor = delegate.tileColor(value)
      numberLabel.textColor = delegate.numberColor(value)
      numberLabel.text = "\(value)"
    }
  }

  unowned let delegate : AppearanceProviderProtocol
  let numberLabel : UILabel

  required init(coder: NSCoder) {
    fatalError("NSCoding not supported")
  }

  init(position: CGPoint, width: CGFloat, value: Int, radius:
      CGFloat, delegate d: AppearanceProviderProtocol) {
    delegate = d
    numberLabel = UILabel(frame: CGRect(x: 0, y: 0, width: width,
        height: width))
    numberLabel.textAlignment = NSTextAlignment.center
    numberLabel.minimumScaleFactor = 0.5
    numberLabel.font = delegate.fontForNumbers()

    super.init(frame: CGRect(x: position.x, y: position.y, width:
        width, height: width))
    addSubview(numberLabel)
    layer.cornerRadius = radius

    self.value = value
    backgroundColor = delegate.tileColor(value)
    numberLabel.textColor = delegate.numberColor(value)
    numberLabel.text = "\(value)"
  }
}
```

图 25.28 砖块视图类 TileView

```
protocol ScoreViewProtocol {
  func scoreChanged(to s: Int)
}
```

图 25.29 ScoreViewProtocol 协议

如图 25.30 所示，计分板视图类遵守计分板视图协议 ScoreViewProtocol 和 UIView，负责游戏计分板的显示。该类定义了计分板的分数属性和显示标签 label，另外还提供了初始化方法和分数改变的函数。

如图 25.31 所示，控制视图类通过常量属性 defaultFrame 定义了视图默认框架的尺寸。

```swift
class ScoreView : UIView, ScoreViewProtocol {
  var score : Int = 0 {
    didSet {
      label.text = "SCORE: \(score)"
    }
  }

  let defaultFrame = CGRect(x: 0, y: 0, width: 140, height: 40)
  var label: UILabel

  init(backgroundColor bgcolor: UIColor, textColor tcolor: UIColor,
      font: UIFont, radius r: CGFloat) {
    label = UILabel(frame: defaultFrame)
    label.textAlignment = NSTextAlignment.center
    super.init(frame: defaultFrame)
    backgroundColor = bgcolor
    label.textColor = tcolor
    label.font = font
    layer.cornerRadius = r
    self.addSubview(label)
  }

  required init(coder aDecoder: NSCoder) {
    fatalError("NSCoding not supported")
  }

  func scoreChanged(to s: Int) {
    score = s
  }
}
```

图 25.30　ScoreView 类

```swift
class ControlView {
  let defaultFrame = CGRect(x: 0, y: 0, width: 140, height: 40)
}
```

图 25.31　ControlView 类

25.2.3　游戏板视图

创建游戏板视图文件 GameboardView.swift。该文件只定义了一个游戏板视图类 GameboardView。如图 25.32 所示，GameboardView 中定义了游戏板相关的属性，包括尺寸、砖块宽度、砖块边框尺寸、边角半径以及砖块视图的集合等变量和砖块移动中的相关常量及默认值。

如图 25.33 所示，在 GameboardView 类中定义了初始化方法 init()，负责对类中的变量属性赋初始值。

```swift
class GameboardView : UIView {
    var dimension: Int
    var tileWidth: CGFloat
    var tilePadding: CGFloat
    var cornerRadius: CGFloat
    var tiles: Dictionary<IndexPath, TileView>
    
    let provider = AppearanceProvider()
    
    let tilePopStartScale: CGFloat = 0.1
    let tilePopMaxScale: CGFloat = 1.1
    let tilePopDelay: TimeInterval = 0.05
    let tileExpandTime: TimeInterval = 0.18
    let tileContractTime: TimeInterval = 0.08
    
    let tileMergeStartScale: CGFloat = 1.0
    let tileMergeExpandTime: TimeInterval = 0.08
    let tileMergeContractTime: TimeInterval = 0.08
    
    let perSquareSlideDuration: TimeInterval = 0.08
```

图 25.32 GameboardView 类

```swift
init(dimension d: Int, tileWidth width: CGFloat, tilePadding
    padding: CGFloat, cornerRadius radius: CGFloat,
    backgroundColor: UIColor, foregroundColor: UIColor) {
    assert(d > 0)
    dimension = d
    tileWidth = width
    tilePadding = padding
    cornerRadius = radius
    tiles = Dictionary()
    let sideLength = padding + CGFloat(dimension)*(width + padding)
    super.init(frame: CGRect(x: 0, y: 0, width: sideLength, height:
        sideLength))
    layer.cornerRadius = radius
    setupBackground(backgroundColor: backgroundColor, tileColor:
        foregroundColor)
}

required init(coder: NSCoder) {
    fatalError("NSCoding not supported")
}
```

图 25.33 GameboardView 的初始化方法

如图 25.34 所示，在 GameboardView 类中定义了 positionIsValid() 函数，用来判断给定的位置是否超出了游戏板的范围。

如图 25.35 所示，在 GameboardView 类中定义了 reset() 函数，用来重置游戏板，即清除游戏板中所有的砖块。

```swift
func positionIsValid(_ pos: (Int, Int)) -> Bool {
  let (x, y) = pos
  return (x >= 0 && x < dimension && y >= 0 && y < dimension)
}
```

图 25.34　GameboardView 的 positionIsValid()函数

```swift
func reset() {
  for (_, tile) in tiles {
    tile.removeFromSuperview()
  }
  tiles.removeAll(keepingCapacity: true)
}
```

图 25.35　GameboardView 的 reset()函数

如图 25.36 所示，在 GameboardView 类中定义了 setupBackground()函数，用来设置游戏板的背景视图。

```swift
func setupBackground(backgroundColor bgColor: UIColor, tileColor:
    UIColor) {
  backgroundColor = bgColor
  var xCursor = tilePadding
  var yCursor: CGFloat
  let bgRadius = (cornerRadius >= 2) ? cornerRadius - 2 : 0
  for _ in 0..<dimension {
    yCursor = tilePadding
    for _ in 0..<dimension {
      // Draw each tile
      let background = UIView(frame: CGRect(x: xCursor, y:
          yCursor, width: tileWidth, height: tileWidth))
      background.layer.cornerRadius = bgRadius
      background.backgroundColor = tileColor
      addSubview(background)
      yCursor += tilePadding + tileWidth
    }
    xCursor += tilePadding + tileWidth
  }
}
```

图 25.36　GameboardView 的 setupBackground()函数

如图 25.37 所示，在 GameboardView 类中定义了 insertTile()函数，用来向游戏板中的特定位置插入一个特定值的砖块。

如图 25.38 所示，在 GameboardView 类中定义了 moveOneTile()函数，用来执行将一个砖块从游戏板的某个位置移到另一个位置的相关操作。

如图 25.39 所示，在 GameboardView 类中定义了 moveTwoTiles()函数，用来执行在游戏板中将两个给定位置的砖块向某个位置移动的相关操作。

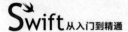

```
func insertTile(at pos: (Int, Int), value: Int) {
  assert(positionIsValid(pos))
  let (row, col) = pos
  let x = tilePadding + CGFloat(col)*(tileWidth + tilePadding)
  let y = tilePadding + CGFloat(row)*(tileWidth + tilePadding)
  let r = (cornerRadius >= 2) ? cornerRadius - 2 : 0
  let tile = TileView(position: CGPoint(x: x, y: y), width:
      tileWidth, value: value, radius: r, delegate: provider)
  tile.layer.setAffineTransform(CGAffineTransform(scaleX:
      tilePopStartScale, y: tilePopStartScale))
  
  addSubview(tile)
  bringSubview(toFront: tile)
  tiles[IndexPath(row: row, section: col)] = tile
  
  UIView.animate(withDuration: tileExpandTime, delay:
      tilePopDelay, options: UIViewAnimationOptions(),
    animations: {
      tile.layer.setAffineTransform(CGAffineTransform(scaleX:
          self.tilePopMaxScale, y: self.tilePopMaxScale))
    },
    completion: { finished in
      UIView.animate(withDuration: self.tileContractTime,
          animations: { () -> Void in
        tile.layer.setAffineTransform(CGAffineTransform.identity)
      })
  })
}
```

图 25.37　GameboardView 的 insertTile() 函数

```
func moveOneTile(from: (Int, Int), to: (Int, Int), value: Int) {
  assert(positionIsValid(from) && positionIsValid(to))
  let (fromRow, fromCol) = from
  let (toRow, toCol) = to
  let fromKey = IndexPath(row: fromRow, section: fromCol)
  let toKey = IndexPath(row: toRow, section: toCol)
  
  guard let tile = tiles[fromKey] else {
    assert(false, "placeholder error")
  }
  let endTile = tiles[toKey]
  
  var finalFrame = tile.frame
  finalFrame.origin.x = tilePadding + CGFloat(toCol)*(tileWidth +
      tilePadding)
  finalFrame.origin.y = tilePadding + CGFloat(toRow)*(tileWidth +
      tilePadding)
  
  tiles.removeValue(forKey: fromKey)
  tiles[toKey] = tile
  
  let shouldPop = endTile != nil
  UIView.animate(withDuration: perSquareSlideDuration,
```

图 25.38　GameboardView 的 moveOneTile() 函数

```
      delay: 0.0,
      options: UIViewAnimationOptions.beginFromCurrentState,
      animations: {
        tile.frame = finalFrame
      },
      completion: { (finished: Bool) -> Void in
        tile.value = value
        endTile?.removeFromSuperview()
        if !shouldPop || !finished {
          return
        }
        tile.layer.setAffineTransform(CGAffineTransform(scaleX:
            self.tileMergeStartScale, y: self.tileMergeStartScale))
        UIView.animate(withDuration: self.tileMergeExpandTime,
          animations: {
            tile.layer.setAffineTransform(CGAffineTransform(scaleX:
                self.tilePopMaxScale, y: self.tilePopMaxScale))
        },
          completion: { finished in
            UIView.animate(withDuration: self.tileMergeContractTime,
              animations: {

                  tile.layer.setAffineTransform(CGAffineTransform
                  .identity)
        })
      })
  })
}
```

图 25.38 （续）

```
func moveTwoTiles(from: ((Int, Int), (Int, Int)), to: (Int, Int),
    value: Int) {
  assert(positionIsValid(from.0) && positionIsValid(from.1) &&
      positionIsValid(to))
  let (fromRowA, fromColA) = from.0
  let (fromRowB, fromColB) = from.1
  let (toRow, toCol) = to
  let fromKeyA = IndexPath(row: fromRowA, section: fromColA)
  let fromKeyB = IndexPath(row: fromRowB, section: fromColB)
  let toKey = IndexPath(row: toRow, section: toCol)

  guard let tileA = tiles[fromKeyA] else {
    assert(false, "placeholder error")
  }
  guard let tileB = tiles[fromKeyB] else {
    assert(false, "placeholder error")
  }

  var finalFrame = tileA.frame
  finalFrame.origin.x = tilePadding + CGFloat(toCol)*(tileWidth +
      tilePadding)
  finalFrame.origin.y = tilePadding + CGFloat(toRow)*(tileWidth +
      tilePadding)
```

图 25.39　GameboardView 的 moveTwoTiles() 函数

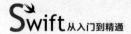

```
    let oldTile = tiles[toKey]
    oldTile?.removeFromSuperview()
    tiles.removeValue(forKey: fromKeyA)
    tiles.removeValue(forKey: fromKeyB)
    tiles[toKey] = tileA

    UIView.animate(withDuration: perSquareSlideDuration,
      delay: 0.0,
      options: UIViewAnimationOptions.beginFromCurrentState,
      animations: {
        // Slide tiles
        tileA.frame = finalFrame
        tileB.frame = finalFrame
      },
      completion: { finished in
        tileA.value = value
        tileB.removeFromSuperview()
        if !finished {
          return
        }
        tileA.layer.setAffineTransform(CGAffineTransform(scaleX:
            self.tileMergeStartScale, y: self.tileMergeStartScale))
        // Pop tile
        UIView.animate(withDuration: self.tileMergeExpandTime,
          animations: {
            tileA.layer.setAffineTransform(CGAffineTransform(scaleX:
                self.tilePopMaxScale, y: self.tilePopMaxScale))
          },
          completion: { finished in
            // Contract tile to original size
            UIView.animate(withDuration: self.tileMergeContractTime,
              animations: {
                tileA.layer.setAffineTransform(CGAffineTransform
                .identity)
            })
        })
    })
  }
}
```

图 25.39 （续）

25.3 控制层

在完成应用的模型层和视图层之后，需要构建负责这两层交互的控制层。

如图 25.40 所示，在项目的主视图控制器 ViewController.swift 中，添加 Button "Start Game" 的动作处理函数 startGameButtonTapped(_ sender: UIButton)。在该函数中创建视图控制器 NumberTileGameViewController 的常量实例 game，并通过

self.present()方法将其显示出来。

```swift
class ViewController: UIViewController {

    override func viewDidLoad() {
        super.viewDidLoad()
    }

    @IBAction func startGameButtonTapped(_ sender : UIButton) {
        let game = NumberTileGameViewController(dimension: 4, threshold: 2048)
        self.present(game, animated: true, completion: nil)
    }
}
```

图 25.40　默认视图控制器

25.3.1　游戏主视图控制器

在 ViewController 中创建 NumberTileGameViewController 的实例。在项目中，创建游戏主视图控制器文件 NumberTileGame.swift，在该文件中定义游戏主视图中所有的视图控制逻辑。

如图 25.41 所示，在文件 NumberTileGame.swift 中定义游戏主视图控制器类 NumberTileGameViewController。在类的声明部分定义游戏板的尺寸属性（游戏板为正方形，边长以砖块数计算）、获胜的阈值属性（默认为 2048）、关联的游戏板视图实例变量、关联的游戏模型实例变量以及计分板视图实例。另外，还设置了游戏板的一系列常量属性，包括宽度、砖块间距、视图间距以及视图垂直缩进等。

```swift
class NumberTileGameViewController : UIViewController,
    GameModelProtocol {
    var dimension: Int
    var threshold: Int

    var board: GameboardView?
    var model: GameModel?

    var scoreView: ScoreViewProtocol?

    let boardWidth: CGFloat = 230.0
    let thinPadding: CGFloat = 3.0
    let thickPadding: CGFloat = 6.0

    let viewPadding: CGFloat = 10.0

    let verticalViewOffset: CGFloat = 0.0
```

图 25.41　游戏主视图控制器类 NumberTileGameViewController

如图25.42所示，在类 NumberTileGameViewController 中继续定义初始化方法。在初始化方法中，对类中的变量属性赋初始值、设置背景色和调用函数 setupSwipeControls()。

```
init(dimension d: Int, threshold t: Int) {
    dimension = d > 2 ? d : 2
    threshold = t > 8 ? t : 8
    super.init(nibName: nil, bundle: nil)
    model = GameModel(dimension: dimension, threshold: threshold,
        delegate: self)
    view.backgroundColor = UIColor.white
    setupSwipeControls()
}

required init(coder aDecoder: NSCoder) {
    fatalError("NSCoding not supported")
}
```

图 25.42　NumberTileGameViewController 的初始化方法

如图 25.43 所示，在类 NumberTileGameViewController 中定义函数 setupSwipeControls()。该函数负责游戏中滑动控件的创建及设置。

```
func setupSwipeControls() {
    let upSwipe = UISwipeGestureRecognizer(target: self, action:
        #selector(NumberTileGameViewController.upCommand(_:)))
    upSwipe.numberOfTouchesRequired = 1
    upSwipe.direction = UISwipeGestureRecognizerDirection.up
    view.addGestureRecognizer(upSwipe)

    let downSwipe = UISwipeGestureRecognizer(target: self, action:
        #selector(NumberTileGameViewController.downCommand(_:)))
    downSwipe.numberOfTouchesRequired = 1
    downSwipe.direction = UISwipeGestureRecognizerDirection.down
    view.addGestureRecognizer(downSwipe)

    let leftSwipe = UISwipeGestureRecognizer(target: self, action:
        #selector(NumberTileGameViewController.leftCommand(_:)))
    leftSwipe.numberOfTouchesRequired = 1
    leftSwipe.direction = UISwipeGestureRecognizerDirection.left
    view.addGestureRecognizer(leftSwipe)

    let rightSwipe = UISwipeGestureRecognizer(target: self, action:
        #selector(NumberTileGameViewController.rightCommand(_:)))
    rightSwipe.numberOfTouchesRequired = 1
    rightSwipe.direction = UISwipeGestureRecognizerDirection.right
    view.addGestureRecognizer(rightSwipe)
}
```

图 25.43　NumberTileGameViewController 的 setupSwipeControls()函数

如图25.44所示，在类 NumberTileGameViewController 中重载视图加载函数

viewDidLoad()和重置函数 reset()。在视图加载函数中调用父类中的 viewDidLoad()函数,然后调用 setupGame()函数创建游戏。另外,还定义了重置函数 reset(),实现游戏中砖块和分数的重置。

```
override func viewDidLoad() {
  super.viewDidLoad()
  setupGame()
}

func reset() {
  assert(board != nil && model != nil)
  let b = board!
  let m = model!
  b.reset()
  m.reset()
  m.insertTileAtRandomLocation(withValue: 2)
  m.insertTileAtRandomLocation(withValue: 2)
}
```

图 25.44 NumberTileGameViewController 的视图载入和重置函数

如图 25.45 所示,在类 NumberTileGameViewController 中定义创建游戏函数 setupGame()。在 viewDidLoad()函数中调用了 setupGame()函数,该函数主要用来对游戏界面进行初始化,包括创建计分板视图 ScoreView 和游戏面板视图 GameboardView。另外,该函数中还定义了两个内嵌函数 xPositionToCenterView()和 yPositionForViewAtPosition()。这两个函数负责计算视图的坐标值 x 和 y,视图根据这个坐标调整位置,从而适应不同尺寸的屏幕。

```
func setupGame() {
  let vcHeight = view.bounds.size.height
  let vcWidth = view.bounds.size.width

  func xPositionToCenterView(_ v: UIView) -> CGFloat {
    let viewWidth = v.bounds.size.width
    let tentativeX = 0.5*(vcWidth - viewWidth)
    return tentativeX >= 0 ? tentativeX : 0
  }
  func yPositionForViewAtPosition(_ order: Int, views: [UIView])
      -> CGFloat {
    assert(views.count > 0)
    assert(order >= 0 && order < views.count)
    let totalHeight = CGFloat(views.count - 1)*viewPadding +
        views.map({ $0.bounds.size.height
        }).reduce(verticalViewOffset, { $0 + $1 })
    let viewsTop = 0.5*(vcHeight - totalHeight) >= 0 ?
        0.5*(vcHeight - totalHeight) : 0
```

图 25.45 NumberTileGameViewController 的 setupGame()函数

```swift
    var acc: CGFloat = 0
    for i in 0..<order {
      acc += viewPadding + views[i].bounds.size.height
    }
    return viewsTop + acc
}

let scoreView = ScoreView(backgroundColor: UIColor.black,
  textColor: UIColor.white,
  font: UIFont(name: "HelveticaNeue-Bold", size: 16.0) ??
      UIFont.systemFont(ofSize: 16.0),
  radius: 6)
scoreView.score = 0

let padding: CGFloat = dimension > 5 ? thinPadding : thickPadding
let v1 = boardWidth - padding*(CGFloat(dimension + 1))
let width: CGFloat =
    CGFloat(floorf(CFloat(v1)))/CGFloat(dimension)
let gameboard = GameboardView(dimension: dimension,
  tileWidth: width,
  tilePadding: padding,
  cornerRadius: 6,
  backgroundColor: UIColor.black,
  foregroundColor: UIColor.darkGray)

let views = [scoreView, gameboard]

var f = scoreView.frame
f.origin.x = xPositionToCenterView(scoreView)
f.origin.y = yPositionForViewAtPosition(0, views: views)
scoreView.frame = f

f = gameboard.frame
f.origin.x = xPositionToCenterView(gameboard)
f.origin.y = yPositionForViewAtPosition(1, views: views)
gameboard.frame = f

view.addSubview(gameboard)
board = gameboard
view.addSubview(scoreView)
self.scoreView = scoreView

assert(model != nil)
let m = model!
m.insertTileAtRandomLocation(withValue: 2)
m.insertTileAtRandomLocation(withValue: 2)
}
```

图 25.45 （续）

如图 25.46 所示，在类 NumberTileGameViewController 中定义函数 followUp()。该函数在每次成功移动砖块后被调用，用来判定游戏是否应该结束。如果结束，则显示结束信息对话框；如果不结束，则在游戏板中的随机位置生成一个砖块。

```swift
func followUp() {
  assert(model != nil)
  let m = model!
  let (userWon, _) = m.userHasWon()
  if userWon {
    let alertView = UIAlertView()
    alertView.title = "Victory"
    alertView.message = "You won!"
    alertView.addButton(withTitle: "Cancel")
    alertView.show()
    return
  }

  let randomVal = Int(arc4random_uniform(10))
  m.insertTileAtRandomLocation(withValue: randomVal == 1 ? 4 : 2)

  if m.userHasLost() {
    NSLog("You lost...")
    let alertView = UIAlertView()
    alertView.title = "Defeat"
    alertView.message = "You lost..."
    alertView.addButton(withTitle: "Cancel")
    alertView.show()
  }
}
```

图 25.46 NumberTileGameViewController 的 followUp() 函数

如图 25.47 所示，在类 NumberTileGameViewController 中添加函数 upCommand() 和 downCommand()。这两个函数负责监听手势识别器消息，当出现向上滑动或向下滑动的手势时，调用 GameModel 中的 queueMove() 函数实现相应的移动操作。

```swift
@objc(up:)
func upCommand(_ r: UIGestureRecognizer!) {
  assert(model != nil)
  let m = model!
  m.queueMove(direction: MoveDirection.up,
    onCompletion: { (changed: Bool) -> () in
      if changed {
        self.followUp()
      }
  })
}
@objc(down:)
func downCommand(_ r: UIGestureRecognizer!) {
  assert(model != nil)
  let m = model!
  m.queueMove(direction: MoveDirection.down,
    onCompletion: { (changed: Bool) -> () in
      if changed {
        self.followUp()
      }
  })
}
```

图 25.47 NumberTileGameViewController 的 upCommand() 和 downCommand() 函数

如图 25.48 所示，在类 NumberTileGameViewController 中添加函数 leftCommand()
和 rightCommand()。这两个函数负责监听手势识别器消息，当出现向左滑动或向右
滑动的手势时，调用 GameModel 中的 queueMove() 函数实现相应的移动操作。

```
@objc(left:)
func leftCommand(_ r: UIGestureRecognizer!) {
  assert(model != nil)
  let m = model!
  m.queueMove(direction: MoveDirection.left,
    onCompletion: { (changed: Bool) -> () in
      if changed {
        self.followUp()
      }
  })
}
@objc(right:)
func rightCommand(_ r: UIGestureRecognizer!) {
  assert(model != nil)
  let m = model!
  m.queueMove(direction: MoveDirection.right,
    onCompletion: { (changed: Bool) -> () in
      if changed {
        self.followUp()
      }
  })
}
```

图 25.48　NumberTileGameViewController 的 leftCommand()和 rightCommand()函数

如图 25.49 所示，在类 NumberTileGameViewController 中添加函数 scoreChanged()。
当分数发生变化时，调用该函数更新计分板显示。添加函数 moveOneTile()，负责将
一个砖块从某个位置移到另一个位置。添加函数 moveTwoTiles()，负责将两个位置
的砖块移到某个位置。添加函数 insertTile()，负责在某个位置插入一个砖块。

```
func scoreChanged(to score: Int) {
  if scoreView == nil {
    return
  }
  let s = scoreView!
  s.scoreChanged(to: score)
}

func moveOneTile(from: (Int, Int), to: (Int, Int), value: Int) {
  assert(board != nil)
  let b = board!
  b.moveOneTile(from: from, to: to, value: value)
}
```

图 25.49　NumberTileGameViewController 的 scoreChanged()、moveOneTile()等函数

```
func moveTwoTiles(from: ((Int, Int), (Int, Int)), to: (Int, Int),
        value: Int) {
    assert(board != nil)
    let b = board!
    b.moveTwoTiles(from: from, to: to, value: value)
}

func insertTile(at location: (Int, Int), withValue value: Int) {
    assert(board != nil)
    let b = board!
    b.insertTile(at: location, value: value)
}
```

图 25.49 （续）

25.3.2 砖块视图的外观定义

在项目中创建文件 AppearanceProvider.swift。该文件负责设置砖块视图的外观，包括游戏中的数字和砖块的外观。该文件由两部分组成：协议 AppearanceProviderProtocol 和类 AppearanceProvider。

如图 25.50 所示，在文件中定义协议 AppearanceProviderProtocol。在该协议中定义了方法 tileColor()、numberColor() 和 fontForNumbers() 的接口。

```
protocol AppearanceProviderProtocol: class {
    func tileColor(_ value: Int) -> UIColor
    func numberColor(_ value: Int) -> UIColor
    func fontForNumbers() -> UIFont
}
```

图 25.50 协议 AppearanceProviderProtocol

如图 25.51 所示，在文件中定义类 AppearanceProvider。该类遵守协议 AppearanceProviderProtocol，并要实现协议接口中定义的全部函数。向类中添加函数 tileColor()，该函数负责设置游戏中砖块的颜色。

```
class AppearanceProvider: AppearanceProviderProtocol {
    func tileColor(_ value: Int) -> UIColor {
        switch value {
        case 2:
            return UIColor(red: 238.0/255.0, green: 228.0/255.0, blue:
                218.0/255.0, alpha: 1.0)
        case 4:
            return UIColor(red: 237.0/255.0, green: 224.0/255.0, blue:
                200.0/255.0, alpha: 1.0)
```

图 25.51 AppearanceProvider 中的 tileColor() 函数

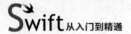

```
case 8:
    return UIColor(red: 242.0/255.0, green: 177.0/255.0, blue:
        121.0/255.0, alpha: 1.0)
case 16:
    return UIColor(red: 245.0/255.0, green: 149.0/255.0, blue:
        99.0/255.0, alpha: 1.0)
case 32:
    return UIColor(red: 246.0/255.0, green: 124.0/255.0, blue:
        95.0/255.0, alpha: 1.0)
case 64:
    return UIColor(red: 246.0/255.0, green: 94.0/255.0, blue:
        59.0/255.0, alpha: 1.0)
case 128, 256, 512, 1024, 2048:
    return UIColor(red: 237.0/255.0, green: 207.0/255.0, blue:
        114.0/255.0, alpha: 1.0)
default:
    return UIColor.white
    }
}
```

图 25.51 （续）

如图 25.52 所示，在类 AppearanceProvider 中添加函数 numberColor（）和 fontForNumbers（）。其中，函数 numberColor（）负责设置游戏中数字的颜色。函数 fontForNumbers（）负责设置游戏中数字的字体。

```
func numberColor(_ value: Int) -> UIColor {
    switch value {
    case 2, 4:
        return UIColor(red: 119.0/255.0, green: 110.0/255.0, blue:
            101.0/255.0, alpha: 1.0)
    default:
        return UIColor.white
    }
}

func fontForNumbers() -> UIFont {
    if let font = UIFont(name: "HelveticaNeue-Bold", size: 20) {
        return font
    }
    return UIFont.systemFont(ofSize: 20)
}
```

图 25.52　AppearanceProvider 中的 numberColor()和 fontForNumbers()函数

25.4 小结

本章介绍了 Apple Store 中的爆款手游 2048 的设计与实现。该游戏的核心代码由 7 个 Swift 文件组成，分别为 GameModel.swift、AuxiliaryModels.swift、AccessoryViews.swift、GameboardView.swift、TileView.swift、NumberTileGame.swift、AppearanceProvider.swift。这 7 个文件又分别属于 MVC 架构中模型层文件、视图层文件和控制层文件。通过这个例子可以看到在几乎不使用可视化控件的情况下，仅通过 Swift 代码就可以设计出复杂有趣的苹果游戏。

第 26 章 编程练习及参考答案

【练习1】 Fibonacci 数列

问题描述：有一对兔子,从出生后第 3 个月起每个月都生一对兔子,小兔子长到第 3 个月后每个月又生一对兔子,假如兔子都不死,每个月的兔子总数为多少?

程序设计思路：先列表找出规律,再根据规律编写程序。具体月份和兔子的对数如下所示。

第 1 个月,只有 1 对兔子。

第 2 个月还是只有 1 对兔子。

第 3 个月开始,这对兔子每个月都会生出一对兔子,当月有 2 对兔子。

第 4 个月增加 1 对兔子,共有 3 对兔子。

第 5 个月开始,第 3 个月出生的一对兔子将开始产兔子,每个月 1 对,当月有 5 对兔子。

依次类推得到,每个月的兔子对数为：1,1,2,3,5,8,13,…。这是一个 Fibonacci 数列,规律就是数列中的前两个数相加得到第 3 个数。

用公式表示就是：$f(n) = f(n-2)+f(n-1)$。

Fibonacci 数列程序及运行结果如图 26.1 所示。在 Playground 中声明一个函数 fibonacciNumbers()。该函数有一个参数为 length,类型为整型,用来描述数列的长度。函数体中定义了 3 个整型变量,用来保存数列中连续的 3 个数,其中第一个数 number1 和第二个数 number2 都初始化为 1,因为第一、二个月的兔子对数均为 1。通过 if 语句判断 length 的长度,如果为 1 或 2,则输出这两个月兔子的对数均为 1。

如果 length 大于 2，则先打印出第一、二个月的兔子对数，然后再通过 for 循环，依次打印出其他月份的兔子对数，具体计算过程是：由 number1 和 number2 计算出 number3，即由前两个月的兔子对数相加得到第 3 个月的兔子对数，然后将 number1 和 number2 的值更新为 number2 和 number3 的值，为下一次计算做好准备。

```
//: Playground - noun: a place where people can play
import UIKit
func fibonacciNumbers(length : Int){
    var number1 = 1                                          1
    var number2 = 1                                          1
    var number3 : Int
    if (length == 1 || length == 2) {
        print("No.\(length) month : 1 pair rabbits")
    }else {
        print("No.1 month : 1 pair rabbits")                 "No.1 month : 1 pair rabbits\n"
        print("No.2 month : 1 pair rabbits")                 "No.2 month : 1 pair rabbits\n"
        for i in 1...length {
            number3 = number1 + number2                      (12 times)
            print("No.\(i) month : \(number3) pair rabbits") (12 times)
            number1 = number2                                (12 times)
            number2 = number3                                (12 times)
        }
    }
}

fibonacciNumbers(12)
```

```
No.1 month : 1 pair rabbits
No.2 month : 1 pair rabbits
No.1 month : 2 pair rabbits
No.2 month : 3 pair rabbits
No.3 month : 5 pair rabbits
No.4 month : 8 pair rabbits
No.5 month : 13 pair rabbits
No.6 month : 21 pair rabbits
No.7 month : 34 pair rabbits
No.8 month : 55 pair rabbits
No.9 month : 89 pair rabbits
No.10 month : 144 pair rabbits
No.11 month : 233 pair rabbits
No.12 month : 377 pair rabbits
```

图 26.1　Fibonacci 数列程序及运行结果

在函数体外调用函数，并传入参数 12。在系统的运行结果输出栏可以看到依次输出了第一个月至第 12 个月的当月兔子总对数，结果正确。

【练习 2】　求质数

问题描述：质数的定义是在大于 1 的自然数中，除 1 和它本身外，不再有其他因数的数。要求计算一个范围内有多少质数，并将其打印出来。

程序设计思路：求一个范围内的质数可以分两步走：第一步判断一个数是否为质数；第二步遍历这个范围，一一判断是否为质数。判断一个数是否为质数可以通过 1 至这个数之间的所有数对其进行整除，如果余数为 0，说明该数有因数，如果没有因数，则说明该数为质数。

求质数程序及运行结果如图 26.2 所示，先实现一个函数 primeNumber()，参数为 number，整型，表示传入待判断的数，返回值为布尔型，为 true 表示该数为质数，反之不是质数。函数体中定义了一个布尔型变量 flag，作为一个是否为质数的标志，初始值为 false。然后通过一个 for 循环，遍历 2 至 number－1 的所有数，将值赋予变量 i。当 number 能够整除 i 的时候，说明 i 为该数的一个因数，并说明该数不为质数，立即停止循环，直接返回 false。如果 number 不能整除 i，则赋值 flag 为 true。循环结束的时候，如果 flag 仍为 true，说明该数不存在因数，返回 true。

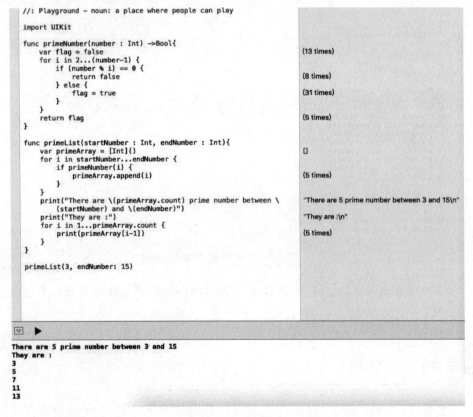

图 26.2　求质数程序及运行结果

然后又定义了一个函数 primeList()，参数为两个整型 startNumber 和 endNumber，分别表示范围的起始值和结束值，没有返回值。在函数体内，首先定义了一个整型数组 primeArray，用来保存所有质数，然后用 for 循环遍历范围内的数，通过调用函数 primeNumber()——判断是否为质数，若为质数，则将其值添加到数组 primeArray 中。最后，通过 primeArray 的长度打印出范围内质数的数量，通过遍历 primeArray 数组打印出范围内所有的质数。

【练习3】 求水仙花数

问题描述：求指定范围内的水仙花数，并打印出来。水仙花数是一个三位数，它的三个组成数字的立方和等于它本身。

程序设计思路：三位数的范围是 100～999，所以首先要遍历给定范围内的所有三位数。其次，需要将三位数的三个数字分解出来，然后进行立方和的运算。

求水仙花数程序及运行结果如图 26.3 所示。定义函数 daffodil()，该函数有两个整型参数 beginNumber 和 endNumber，无返回值。在函数体中，先定义一个整形数组 daffodilArray，用来保存找到的水仙花数。然后，用 for 循环遍历范围内的全部三位数，将其赋值给 i。通过整除和取模两种运算，分别提取该三位数的个、十、百位的数字，并分别保存到常量 firstNumber、secondNumber、thirdNumber。计算这三个常量的立方和，并赋值给常量 cubeNumber。通过比较 cubeNumber 和 i，判断是否为水仙

```
//: Playground - noun: a place where people can play
import UIKit

func daffodil(beginNumber : Int, endNumber : Int){
    var daffodilArray = [Int]()
    for i in beginNumber...endNumber {
        let firstNumber = i / 100
        let secondNumber = i / 10 % 10
        let thirdNumber = i % 10
        let cubeNumber = firstNumber * firstNumber * firstNumber +
            secondNumber * secondNumber * secondNumber + thirdNumber
            * thirdNumber * thirdNumber
        if cubeNumber == i {
            daffodilArray.append(i)
        }
    }
    print("There are \(daffodilArray.count) daffodil number between
        \(beginNumber) and \(endNumber)")

    for i in 1...daffodilArray.count {
        print(daffodilArray[i-1])
    }
}

daffodil(130, endNumber: 166)
```

```
[]

(37 times)
(37 times)
(37 times)
(37 times)




[153]


"There are 1 daffodil number between 130 and 166\n"

"153\n"
```

图 26.3 求水仙花数程序及运行结果

花数，若是，则将其添加到 daffodilArray 数组中。最后，打印水仙花数的个数，并用循环依次读出给定范围内求得的水仙花数。

【练习 4】 统计字符串中的各类字符个数

问题描述：给定一个字符串，统计字符串中大小写字母、数字、空格及其他未识别类型字符的个数。

程序设计思路：遍历字符串，比较字符串中每个字符和各类字符的范围值。

统计字符串中各类字符个数的程序及运行结果如图 26.4 所示。定义函数 judgeAString()，该函数有一个字符串型参数 theString，传入需要进行统计的字符串。函数体中定义了 5 个整型变量，初始化都为 0，分别为 lowerCase 记录小写字符的个数、upperCase 记录大写字符的个数、digital 记录数字的个数、space 记录空格的个数、others 记录其他类型的个数。通过一个 for 循环及 String 类型的 characters 属性，遍历字符串 theString 中的每个字符，将其保存到 i 中。然后用连续的 if-else 语句来分别比较 i 和各类型字符的边界值，从而统计各类型字符的个数。最后，将统计结果打印出来。

```
//: Playground - noun: a place where people can play
import UIKit
func judgeAString(theString : String) {
    var lowerCase = 0                                       0
    var upperCase = 0                                       0
    var digital = 0                                         0
    var space = 0                                           0
    var others = 0                                          0
    for i in theString.characters {
        if ((i >= "a") && (i <= "z")) {
            lowerCase = lowerCase + 1                       (10 times)
        } else if ((i >= "A") && ( i <= "Z")) {
            upperCase = upperCase + 1                       (2 times)
        } else if ( i == " ") {
            space = space + 1                               (4 times)
        } else if ( (i >= "0") && (i <= "9")) {
            digital = digital + 1                           (3 times)
        } else {
            others = others + 1                             (2 times)
        }
    }
    print("There are lower case \(lowerCase), upper case \    "There are lower case 10, upper case 2, space 4, digital 3, others 2\n"
        (upperCase), space \(space), digital \(digital), others \
        (others)")
}
judgeAString("5 2 1, Beautiful Day!")
```

图 26.4 统计字符串中各类字符个数的程序及运行结果

【练习 5】 给定项数和数的和

问题描述：求给定项数的数字 a 的和，即 sum＝a＋aa＋aaa＋aaaa＋aa…a 的值。

程序设计思路：求给定项目的数字的和，其中项数 n 和数字 a 是外部输入的，也就是说这两项是给定的。那么，求和可以分两步走，首先求每一项的值，例如第 3 项的值 aaa 等于 a＋a＊10＋a＊100；其次，就是将 n 项进行相加。

给定项数和数的和的程序及运行结果如图 26.5 所示。定义函数 sumOfA()，该函数有两个整型参数 a 和 numOfTerms，用来表示数字 a 和项数。函数体内定义了 4 个变量，分别为：变量 curTerm 用来表示求和中的每一项数，初始值为 0；变量 sum 用来表示求和的结果，初始值为 0；变量 i 用来控制循环的次数，初始值为 0；变量 fractor 用来表示相邻两项之间的增量，如 2 和 22，增量就为 20，用 a 可以表示为 a＊10。fractor 的初始值为 a。

```
//: Playground - noun: a place where people can play
import UIKit
func sumOfA(a : Int, numOfTerms : Int){
    var curTerm = 0                              0
    var sum = 0                                  0
    var i = 0                                    0
    var fractor = a                              2
    while i < numOfTerms {                       (3 times)
        curTerm = curTerm + fractor              (3 times)
        sum = sum + curTerm                      (3 times)
        fractor = fractor * 10                   (3 times)
        i = i + 1
    }
    print("Sum of \(numOfTerms) a-\(a) is \(sum)")   "Sum of 3 a-2 is 246\n"
}
sumOfA(2, numOfTerms: 3)
```

图 26.5　给定项数和数的和的程序及运行结果

函数体内用一个 while 循环，循环执行 numOfTerms 次数，也就是将 numOfTerms 项的数累加起来。在循环体内，先计算当前项 curTerm。当前项由前一项的值加上增量 fractor 得到。累加和 sum 等于上一次的 sum 值加上当前项的值。每次循环都要将 fractor 的值乘以 10，因为相邻项之间的增量差 10 倍。最后，打印计算结果。

【练习6】 自由落体反弹问题

问题描述：一个球从高处 h 米自由落体，每次落地后都能反弹回上一次高度的一半，求第 n 次反弹的高度，以及一共反弹的米数。

程序设计思路：自由落体反弹问题可以通过循环计算解决。

自由落体反弹问题的程序及运行结果如图 26.6 所示。定义函数 bounceUpCompute()，该函数有两个参数 h 和 n，分别表示高处的米数和反弹的次数。在函数体内定义三个变量：sum、theHeight 和 i。sum 表示反弹的总米数，为双精度浮点数，初始值为 0.0。theHeight 为每次反弹的高度，初始值为 h。i 用来控制循环的次数，初始值为 0。函数体内非常简单，首先计算每次反弹的高度 theHeight，其次累加总的反弹高度，然后循环次数 i 加 1，最后打印出相关的信息。

```
//: Playground - noun: a place where people can play
import UIKit

func bounceUpCompute(h : Double, n : Int) {
    var sum = 0.0
    var theHeight = h
    var i = 0
    while i < n {
        theHeight = theHeight / 2
        sum = sum + theHeight
        i = i + 1
    }
    print("No.\(n) bounce-up height is \(theHeight)m")
    print("The sum of bounce-up distance is \(sum)m")
}

bounceUpCompute(100, n: 10)
```

```
0
100
0
(10 times)
(10 times)
(10 times)
"No.10 bounce-up height is 0.09765625m\n"
"The sum of bounce-up distance is 99.90234375m\n"
```

图 26.6　自由落体反弹问题的程序及运行结果

【练习7】 求无重复的三位数

问题描述：用 1～6 的数字可以组成多少个不同且不含重复数字的三位数。

程序设计思路：从 1～6 的 6 个数字中，取出 3 个数字组合成三位数，这 3 个数字不能重复。可以通过三层循环，穷举各种组合情况，并通过判断语句剔除重复的情况。

求无重复的三位数的程序及运行结果如图 26.7 所示。定义两个变量 theNumber

和 theArray。theNumber 为整型变量,临时保存符合要求的数。theArray 为整型数组,用来保存所有符合要求的数。通过三层循环,遍历 1~6 中的数字组合,将百位、个位及十位分别保存到 i,j,k 中,并进行比较。当三者均不相等时,求得该三位数,并将其添加到数组 theArray 中。然后打印出相应的信息。这里需要注意的是,由于符合要求的数很多,所以直接使用 print 进行打印,一行只能打印出一个数,很不直观。这里考虑一行打印 15 个数,那么就需要在遍历数组 theArray 的时候,再做一个计数控制,即每当计数到 15 的时候,输出这 15 个数组成的字符串。具体实现是:定义一个变量 j,用来计数。定义一个字符串变量 numberString,用来拼接一次性打印的 15 个数。通过 for 循环遍历数组,并将其中的数取出赋值到 i 中。将 i 转换成字符串,赋值

```
//: Playground - noun: a place where people can play
import UIKit
func threeBitsNumber(){
    var theNumber : Int
    var theArray = [Int]()
    for i in 1...6 {
        for j in 1...6 {
            for k in 1...6 {
                if ((i != j)&&(i != k)&&(j != k)) {
                    theNumber = i * 100 + j * 10 + k
                    theArray.append(theNumber)
                }
            }
        }
    }
    print("There are \(theArray.count) different numbers. They are : ")
    var j = 1
    var numberString = ""
    for i in theArray {
        numberString = numberString + String(i) + " "
        if j == 15 {
            print(numberString)
            numberString = ""
            j = 0
        }
        j = j + 1
    }
}

threeBitsNumber()
```

```
There are 120 different numbers. They are :
123 124 125 126 132 134 135 136 142 143 145 146 152 153 154
156 162 163 164 165 213 214 215 216 231 234 235 236 241 243
245 246 251 253 254 256 261 263 264 265 312 314 315 316 321
324 325 326 341 342 345 346 351 352 354 356 361 362 364 365
412 413 415 416 421 423 425 426 431 432 435 436 451 452 453
456 461 462 463 465 512 513 514 516 521 523 524 526 531 532
534 536 541 542 543 546 561 562 563 564 612 613 614 615 621
623 624 625 631 632 634 635 641 642 643 645 651 652 653 654
```

图 26.7 求无重复的三位数的程序及运行结果

给 numberString。再通过 if 语句判断 j 是否达到 15，若达到，则打印 numberString，并重置 numberString 为空，重置 j 为 0。在 if 语句外，j 赋值为 j+1，然后进入下一轮循环。

【练习8】 阶梯奖金计算

问题描述：企业根据个人产生的利润进行提成，从而确定发放给个人的奖金额度。具体规则为：

利润低于或等于 10 万元时，奖金可从中提成 10%。

利润高于 10 万元，且低于或等于 20 万元时，高于 10 万元的部分可提成 7.5%，其他部分按前面规则提取。

利润高于 20 万，且低于或等于 40 万元时，高于 20 万元的部分可提成 5%，其他部分按前面规则提取。

利润高于 40 万，且低于或等于 60 万元时，高于 40 万元的部分可提成 3%，其他部分按前面规则提取。

利润高于 60 万，且低于或等于 100 万元时，高于 60 万元的部分可提成 1.5%，其他部分按前面规则提取。

利润高于 100 万元时，超过 100 万元的部分可提成 1%，其他部分按前面规则提取。

要求根据给定的利润计算应发的奖金。

程序设计思路：根据利润所属的不同分档区间分别进行计算。既可以从最低档开始计算，也可以从最高档开始计算。这里采用由高到低的方法，即先判断利润属于哪个档次，然后计算该档次的奖金，该档次以下的部分按照最高值计算即可。

阶梯奖金计算的程序及运行结果如图 26.8 所示。定义函数 arrangeBonus()，该函数有一个参数 profits，为双精度浮点型，表示需要计算奖金的利润数。在函数体内，定义了两个浮点型变量，bonus 表示奖金数，初始值为 0.0；theProfits 表示带计算的利润数，初始值为 profits。然后，通过一系列的 if 语句匹配当前带计算的利润数。当匹配一个利润值的区间后，首先用当前待计算的利润数 theProfits 减掉该区间的下限值，多出的部分按照该区间的提成比率计算，计算结果累加到 bonus 中。然后将当

```
//: Playground - noun: a place where people can play
import UIKit
func arrangeBonus(profits : Double){
    var bonus = 0.0                                     0
    var theProfits = profits                            135
    if (theProfits > 100) {
        bonus += (theProfits - 100) * 0.01              0.35
        theProfits = 100                                100
    }
    if ((theProfits > 60) && (theProfits <= 100)) {
        bonus += (theProfits - 60) * 0.015              0.95
        theProfits = 60                                 60
    }
    if ((theProfits > 40) && (theProfits <= 60)) {
        bonus += (theProfits - 40) * 0.03               1.55
        theProfits = 40                                 40
    }
    if ((theProfits > 20) && (theProfits <= 40)) {
        bonus += (theProfits - 20) * 0.05               2.55
        theProfits = 20                                 20
    }
    if ((theProfits > 10) && (theProfits <= 20)) {
        bonus += (theProfits - 10) * 0.075              3.3
        theProfits = 10                                 10
    }
    if ((theProfits > 0) && (theProfits <= 10)) {
        bonus += theProfits * 0.1                       4.3
    }
    print("Profits is \(theProfits)")                   "Profits is 10.0\n"
    print("Bonus is \(bonus)")                          "Bonus is 4.3\n"
}
arrangeBonus(135)
```

图 26.8　阶梯奖金计算的程序及运行结果

前待计算的利润数 theProfits 调整为该区间的下限值，继续匹配其他情况。在运行过程中，实际传入的利润数为 135。从第一条 if 语句开始匹配，计算得到奖金 0.35。然后第二条 if 语句匹配后，得到累加奖金 0.95。依次执行完语句，最后得到的奖金数为 4.3。如果输入的利润数为 8，则只会执行函数体中的最后一个 if 语句，其他语句都不匹配。

【练习 9】　求完全平方数

问题描述：求一定范围内的数，它满足加上 100 和 268 均为完全平方数。完全平方数就是一个数正好是别的数的平方。

程序设计思路：判断一个数是否为平方数，可以通过对其开方后，再对 1 取模，如果为 0，说明正好是一个数的平方，开方后没有小数部分。

求完全平方数的程序及运行结果如图 26.9 所示。定义函数 squareNumber()，该函数含有一个参数 scope，为浮点型，表示计算的范围。定义两个浮点型变量 r1 和 r2，用来保存计算结果。通过 while 循环，遍历范围内的数，并将其赋值给 i。在循环体内计算 i+100，然后开方，再模 1，将值保存到 r1。同样，计算 i+268，然后开方，再模 1，将值保存到 r2。通过 if 语句判断，r1 和 r2 是否同时为 0，如果是，则打印出这个数，否则 i 加 1，继续循环。

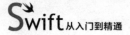

图 26.9　求完全平方数的程序及运行结果

【练习10】　求一年中的第几天

问题描述：根据给定的年、月、日信息，判断这一天为一年中的第几天。

程序设计思路：根据年、月、日计算某一天属于一年中的第几天，需要注意两个问题：第一，根据年份确定该年是否为闰年。闰年是指能被 400 整除或者能被 4 整除但不能被 100 整除的年份。闰年的特点是该年的 2 月为 29 天。第二，根据月份计算已经过去的几个月中的天数和，这里需要注意月份要减 1，因为当月的天数由天数决定。下面就是如何计算，可以通过 for 循环，对给定的月份进行遍历，月份的天数可以通过 switch 语句进行匹配。

求一年中的第几天的程序及运行结果如图 26.10 所示。定义函数 theDayOfTheYear()，该函数有 3 个参数 year、month、day，为整型，分别表示年、月、日，返回值为整型，表示在一年中的天数。在函数体内定义两个变量，leapYear 为布尔型，表示是否为闰年，theDays 为整型，表示在一年中的天数，初始值为 day，即传入的天数。首先判断月份和天数的信息是否为一个合法的值，如果不是，则打印异常信息。然后判断年份是否为闰年，若是，则赋值 leapYear 为 true。通过 for 循环遍历月份，通过 switch 语句匹配月份和天数，最后通过累加得到天数 theDays。运行结果显示，2000 年和 2001 年的同月同日在一年中的天数相差一天，原因是 2000 年为闰年。

```
//: Playground - noun: a place where people can play
import UIKit
func theDayOfTheYear(year : Int, month : Int, day : Int) -> Int {    (2 times)
    var leapYear = false                                              (2 times)
    var theDays = day
    if (month > 12) || (month < 1) {
        print("month is illegal")
        return 0
    }
    if (day > 31) || ( day < 1) {
        print("day is illegal")
        return 0
    }
    if ((year % 400 == 0) || ((year % 4 == 0) && (year % 100 == 0))) {
        leapYear = true                                               true
    }
    for i in 1...month-1 {
        switch i {
        case 1,3,5,7,8,10,12 : theDays += 31                          (4 times)
        case 4,6,9,11 : theDays += 30                                 (2 times)
        case 2 : if leapYear {
            theDays += 29                                             76
        } else {
            theDays += 28                                             75
        }
        default : print("It's error")
        }
    }
    return theDays                                                    (2 times)
}
theDayOfTheYear(2000, month: 5, day: 16)                              137
theDayOfTheYear(2001, month: 5, day: 16)                              136
```

图 26.10　求一年中的第几天的程序及运行结果

【练习 11】　3 个数比大小

问题描述：将任意 3 个数按照从大到小的顺序输出。

程序设计思路：比较 3 个数的大小，并按照大小顺序输出，这是一个比较简单的排序，通过几个 if 语句逐个进行比较即可实现。

3 个数比大小的程序及运行结果如图 26.11 所示。定义函数 compare3Numbers()，该函数有 3 个参数 x、y、z，分别接收外部输入的 3 个待排序的整数。在函数体中，定义 3 个整型变量 big、middle、small，初始值都为 0，用来存储 3 个数中的最大数、中间数、最小数。通过 if 语句判断 x 和 y 的大小，并根据大小关系给 big 和 middle 赋值，暂时不考虑 small 的值，然后比较 z 和 big、middle 的关系。具体有 3 种情况：z>big、middle<z<big、z<middle。根据这 3 种情况，重新为 big、middle、small 赋值，然后按照顺序将 3 个数输出。

```
//: Playground - noun: a place where people can play
import UIKit

func compare3Numbers(x : Int, y : Int, z : Int) {
    var big = 0, middle = 0, small = 0
    if x > y {
        big = x
        middle = y
    }else{
        big = y
        middle = x
    }
    
    if z > big {
        small = middle
        middle =  big
        big = z
    }else if ((z < big) && (z > middle)) {
        small = middle
        middle = z
    }else if z < middle {
        small = z
    }
    
    print("Numbers from big to small is \(big)  \(middle)   \(small)")
}
compare3Numbers(6, y: 8, z: 2)
compare3Numbers(9, y: 3, z: 12)
compare3Numbers(1, y: 2, z: 3)
```

```
Numbers from big to small is 8  6  2
Numbers from big to small is 12  9  3
Numbers from big to small is 3  2  1
```

图 26.11　3 个数比大小的程序及运行结果

【练习 12】 打印九九乘法表

问题描述：打印九九乘法表，形式为：a＊b＝c。

程序设计思路：九九乘法表的计算非常简单，主要问题在于打印。九九乘法表的规律是一共 9 行 9 列，每行的被乘数相同，每列的乘数也相同。所以，只通过两层循环遍历 1～9 的数字即可。

打印九九乘法表的程序及运行结果如图 26.12 所示。该定义函数 multiplyTable()，该函数无参数和返回值。函数体内定义了整型变量 result，用来记录运算结果；定义了字符串变量 lineString，初始值为空，用来存储乘法表的每一行字符串。通过两层 for 循环，依次从 1～9 进行遍历，计算并打印出各表达式及运算结果。最后调用该函数打印出九九乘法表，并输出到控制台，可以看到符合我们的预期设计要求。

```
//: Playground - noun: a place where people can play
import UIKit

func multiplyTable(){
    var result : Int
    var lineString = ""
    for i in 1...9 {
        for j in 1...9 {
            result = i * j
            if result < 10 {
                lineString += "\(i)*\(j)=\(result)   "
            } else {
                lineString += "\(i)*\(j)=\(result)  "
            }
        }
        print(lineString)
        lineString = ""
    }
}

multiplyTable()
```

图 26.12 打印九九乘法表的程序及运行结果

【练习 13】 猴子吃桃问题

问题描述：猴子摘下了若干个果子，第一天吃了一半后，不过瘾，又多吃了一个，第二天吃了剩下的一半后，不过瘾，又多吃了一个。此后，每天如此，均吃了剩下的一半多一个果子。到某一天时，发现只剩下一个果子。编程实现：告诉你第几天只剩下一个果子，求总共摘了多少果子？

程序设计思路：到哪一天只剩下一个果子，天数是一个变数，直接决定了总共的果子数，这里是作为参数由外部输入的。那么，如何在给定的天数下计算总的果子数呢？可以采用倒推的方法，最后一天剩下一个果子，那么前一天，吃了一半后，又多吃了一个果子，由此可以判定最后一个果子加上多吃的一个果子等于前一天的一半。由此，可以通过循环反推果子的总数，而循环的次数则取决于一共吃了多少天。

猴子吃桃问题的程序及运行结果如图 26.13 所示。定义函数 monkeyEatPeach()，该函数有一个参数 days，为整型，由函数调用者输入，表示天数，即经过多少天，达到只剩下一个果子的状态。在函数体中定义一个变量 peachNum，用来保存当前的桃子数，初始值为 1，即表示最后一天只剩下一个桃子。通过 for 循环执行 days－1 次，因为最后一次果子数为 1，已经赋值给 peachNum 了。在循环体里，通过式（peachNum＋1）*2 计算当天果子数。循环结束的时候，可以得到第一天的果子数，即一共有多少桃子。

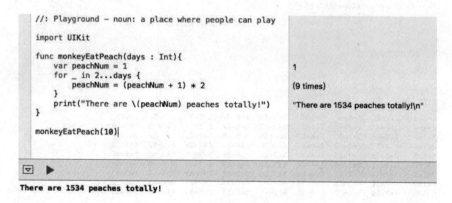

图 26.13 猴子吃桃问题的程序及运行结果

【练习 14】　求分数数列的和

问题描述：有一个分数数列：2/1、3/2、5/3、8/5、13/8、21/13，…，求这个分数数列前 n 项的和。

程序设计思路：首先需要分析这个分数数列的规律，然后求出每一项，通过循环将前 n 项的分数进行累加。从第二个分数开始，每一个分数的分母都是前一个分数的分子，分子是前一个分数的分子与分母的和。

求分数数列的和的程序及运行结果如图 26.14 所示。定义函数 sumOfFractionSeries()，该函数有一个整型参数 n，用来输入分数数列的长度。函数体中定义了 5 个变量：变量 numerator 表示分子部分，初始值为第一个分数的分子值 2.0；变量 denominator 表示分母部分，初始值为第一个分数的分母值 1.0；变量 temp 赋值为 0.0，用来作为临时保存数的地方；变量 sum 为分数累加的和，初始值为第一个分数的值 2.0；变量 resultStr 为字符串型，用来保存要打印输出的结果，初始值为第一个分数的字符串 "2/1"。通过 for 循环，对余下的 n－1 个分数进行累加。在循环体内，先根据前一个分数的分子和分母求得当前的分子和分母的值，然后将新分数的值累加到 sum 中。另外，还需要在输出的字符串中连接上新的分数的字符串。循环结束后，打印分数数列和计算结果。

```
//: Playground - noun: a place where people can play

import UIKit

func sumOfFractionSeries(n : Int) {
    var numerator = 2.0                                              2
    var denominator = 1.0                                            1
    var temp = 0.0                                                   0
    var sum = 2.0                                                    2
    var resultStr = "2/1"                                            "2/1"
    for _ in 2...n {
        temp = numerator                                             (7 times)
        numerator = denominator + numerator                          (7 times)
        denominator = temp                                           (7 times)
        sum += numerator/denominator                                 (7 times)
        resultStr += " + " + String(Int(numerator)) + "/" + String(Int   (7 times)
            (denominator))
    }
    print(resultStr + " = \(sum)")                                   "2/1 + 3/2 + 5/3 + 8/5 +
}

sumOfFractionSeries(8)

2/1 + 3/2 + 5/3 + 8/5 + 13/8 + 21/13 + 34/21 + 55/34 = 13.2437459599224
```

图 26.14　求分数数列的和的程序及运行结果

【练习 15】 求 n 的阶乘的和

问题描述：求 1!＋2!＋…＋n!的和。

程序设计思路：一个正整数的阶乘指的是所有小于或等于这个数的正整数的乘积。求阶乘的和，首先要求出每项阶乘的值，然后进行累加。

求 n 的阶乘的和的程序及运行结果如图 26.15 所示。定义 sumOfFactorial()，该函数还有一个整型变量 n，用来表示计算阶乘的最后一个数。函数体中定义了两个整型变量 sum 和 f，一个字符串变量 str。变量 sum 用来表示阶乘的和，初始值为 1!，即 1。变量 f 表示当前的阶乘的值，初始值为 1!，即 1。变量 str 用来打印阶乘相加的字符串，初始值为第 1 个阶乘 1!。通过 for 循环从第 2 个阶乘开始，首先计算阶乘的值，然后累加，最后更新 str 的值。在循环体外，打印出 n 个阶乘相加的结果。

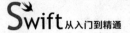

图 26.15　求 n 的阶乘的和的程序及运行结果

【练习 16】 用递归法求阶乘

问题描述：阶乘的定义在练习 15 已经给出，这里需要用递归的方法求阶乘。所谓递归，就是在运行的过程中调用自己。构成递归有两个条件：首先，子问题和原问题是相同的事情,且更加简单;其次,不能无限制地调用本身,必须有一个递归出口。

程序设计思路:计算阶乘的方法还沿用练习 15 的方法。这里只需要定义一个递归函数,给出求某一项阶乘的算法,即 n!可以转化为(n−1)!,依次往前推。递归的出口在于 1!=1。

用递归法求阶乘的程序及运行结果如图 26.16 所示。定义递归函数 factorial(),该函数有一个整型参数 n,用来计算它的阶乘,返回值为整型,即阶乘的值。函数体里定义了一个整型变量 result,初始值为 0,用来保存阶乘的值。当 n 为 1 的时候,直接得出阶乘的值,赋值 result 等于 1,当 n 不为 1 的时候(阶乘是指正整数连乘的积,不考虑负数的情况,所以 n 不为 1 相当于 n 大于 1),result 等于 n 乘以 n−1 的阶乘,这里 n−1 的阶乘是通过调用函数本身实现的。这个调用过程会一直向前递归,直到递归出口条件,即 n−1=1 的时候,然后再逐层往回迭代。递归的含义也正在此,即先将 n−1 一层层"递"出去,直到递归出口条件 1 为止,然后再一层层"归"回来。

```
//: Playground - noun: a place where people can play
import UIKit

func factorial(n : Int) -> Int {
    var result = 0                          (15 times)
    if n == 1 {
        result = 1                          (5 times)
    }else{
        result = n * factorial(n-1)         (10 times)
    }
    return result                           (15 times)
}

print("1! = " + String(factorial(1)))      "1! = 1\n"
print("2! = " + String(factorial(2)))      "2! = 2\n"
print("3! = " + String(factorial(3)))      "3! = 6\n"
print("4! = " + String(factorial(4)))      "4! = 24\n"
print("5! = " + String(factorial(5)))      "5! = 120\n"
```

```
1! = 1
2! = 2
3! = 6
4! = 24
5! = 120
```

图 26.16 用递归法求阶乘的程序及运行结果

【练习 17】 倒推年龄

问题描述:有 n 个人坐在一起,问第 n 个人多少岁,他说比第 n−1 个人大 2 岁。问第 n−1 个人多少岁,他说比第 n−2 个人大 2 岁。接着问第 n−3 个人,一直问到第一个人，第一个人说他今年 m 岁。请根据 n 和 m 的值，计算出第 n 个人的岁数。

程序设计思路：本题的关键是第 n 个人与第 n-1 个人之间的年龄差总是 2 岁。那么，从第一个人的岁数开始，通过不断加 2 的方式，就能不断往前推断前面人的岁数，一直推断到第 n 个人的岁数。

倒推年龄的程序及运行结果如图 26.17 所示。定义函数 ageProblem()，该函数有两个参数 age 和 n，均为整型，分别表示第一个人的年龄和一共有多少个人。在函数体中定义一个整型变量 lastOne，用来保存第 n 个人的年龄，初始值为 age，即第一个人的年龄。通过 for 循环从第二个人开始循环，一直到第 n 个人为止。每次循环都对 lastOne 递增 2。计算完毕后，在循环体外打印第 n 个人的年龄信息。

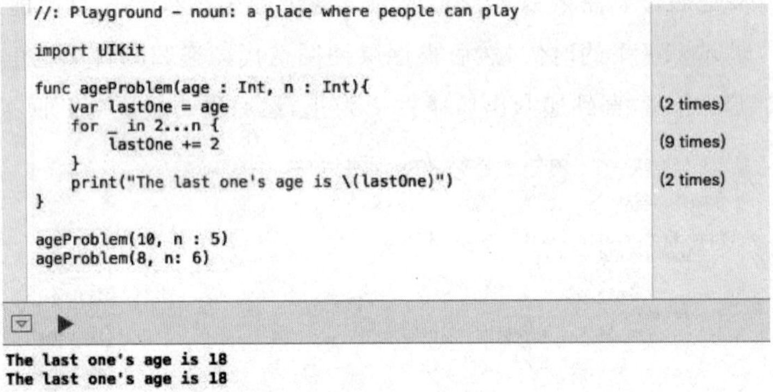

图 26.17　倒推年龄的程序及运行结果

【练习 18】　逆序打印一个整数

问题描述：给定一个正整数，并逆序打印出来。

程序设计思路：要逆序打印一个正整数，需要分两步走：第一步，将这个正整数转换为一个字符串；第二步，遍历这个字符串，将其出现的顺序调整一下。

逆序打印一个整数的程序及运行结果如图 26.18 所示。定义函数 numberReversed()，该函数有一个 64 位整型参数 number，之所以用 64 位整型，就是为了能够处理较大的整数。函数体内定义了两个变量：变量 numStr 为字符串型，用来存储参数 number 的字符串形式；变量 reversedStr 是字符串型，初始值为空，用来保存逆序排列的整数的字符串形式。首先，将整数 number 转换为字符串形式，并保存在变量 numStr 中。

其次，通过 for 循环依次遍历 numStr 中的字符，并将其取出赋值给 i。然后，由字符 i 和 reversedStr 拼接出新的字符串。需要注意的是，每个新取出的 i 都要添加在字符串 reversedStr 的前面，这样就可以实现逆序排列了。最后，在循环体外打印出输入的整数和逆序排列的整数的信息。

```
//: Playground - noun: a place where people can play
import UIKit

func numberReversed(number : Int64){
    var numStr : String
    var reversedStr = ""
    numStr = String(number)
    for i in numStr.characters {
        reversedStr = String(i) + reversedStr
    }
    print("The input string is \(numStr)")
    print("The reversed string is " + reversedStr)
}

numberReversed(12345)
numberReversed(67890)
```

```
The input string is 12345
The reversed string is 54321
The input string is 67890
The reversed string is 09876
```

图 26.18　逆序打印一个整数的程序及运行结果

【练习 19】　回文问题

问题描述：回文就是正着读和反着读都一样的字符串。给定一个整数，判断该整数是否为回文。

程序设计思路：回文是指符合条件的字符串，所以首先要将整数转换为字符串。为了对字符串里的每个字符进行处理，所以将字符串转换为字符数组。通过字符数组判断是否符合回文的特点，假定数组的长度为 n，实际就是比较数组的第 i 个元素和第 n−i 个元素是否相等。结合数组类型的方法 first()、last()、removeFirst()、removeLast()，可以每次都比较该数组的第一个元素和最后一个元素是否相等，如果不相等，则说明不是回文，如果相等，则从数组中去掉第一个元素和最后一个元素，然后继续判断第一个元素和最后一个元素是否相等。

回文问题的程序及运行结果如图 26.19 所示。定义函数 JudgePalindrome()，该函数有一个 64 位整型参数，用来输入待判断的整数。在函数体内，首先定义了字符串常量 str，该常量的初始值为输入整型参数 number 的字符串值；定义整型常量 halfCount，用来记录字符串长度的一半。例如，字符串长度为 5，其长度一般算作 2；定义字符型数组 charArray；定义布尔型变量 palindrome，初始值为 true，用来标识该数是否为回数，默认为回数，下面的程序将对其进行判断，当发现有不符合回数的情况，则该值就会被赋值 false。如果检查完以后，变量 palindrome 仍然为 true，就说明该数为回数。通过 for 循环遍历 str.characters，将其中的字符逐个添加到字符型数组 charArray 中。通过 for 循环，执行 halfCount 次，每次都比较 charArray 的第一个字符和最后一个字符，如果不相等，则赋值 palindrome 为 false，并跳出循环。如果相等，则从数组 charArray 中去掉第一个和最后一个字符，并判断此时的数组长度是否为 1，若为 1，则说明已经到中间的那个数，此时应该跳出循环，若不为 1，则继续下一轮循环。最后，通过两个整数 12345 和 12321 测试函数的正确性，结果在右侧同步输出。

```swift
//: Playground - noun: a place where people can play

func JudgePalindrome(number : Int64){
    
    let str = String(number)                              (2 times)
    let halfCount = str.characters.count / 2              (2 times)
    var charArray = [Character]()                         (2 times)
    var palindrome = true                                 (2 times)
    
    for char in str.characters {
        charArray.append(char)                            (10 times)
    }
    
    for _ in 1...halfCount {
        if (charArray.first != charArray.last) {
            palindrome = false                            false
            break
        }
        charArray.removeFirst()                           (2 times)
        charArray.removeLast()                            (2 times)
        if charArray.count == 1 {
            break
        }
    }
    
    if palindrome {
        print("Integer \(number) is a palindrome!")       "Integer 12321 is a palindrome!\n"
    } else {
        print("Integer \(number) is not a palindrome.")   "Integer 12345 is not a palindrome.\n"
    }
    
}

JudgePalindrome(12345)
JudgePalindrome(12321)
```

图 26.19　回文问题的程序及运行结果

【练习 20】 整数排序

问题描述：给定一个整型数组，对数组中的元素按照从小到大的顺序排序。

程序设计思路：可以将数组排序分解为相邻的两个数进行比较，如果前一个数小于后一个数，则交换两者在数组中的位置。通过两层 for 循环，可以两两比较数组中的所有元素。

整数排序的程序及运行结果如图 26.20 所示。定义函数 sortNumberArray()，该函数含有一个整型数组型的变量 numbers，用来接收外部传入的待排序整型数组，返回值也为一个整型数组，用来返回排好序的整型数组。函数体内定义了整型数组变量 sortedNumbers，初始值为未排序的整型数组，将来存储排序后的数组；定义了常量 count，初始值为数组长度减 1，用来控制循环的次数；定义了变量 temp，初始值为 0，用来作为中转用的整型变量。通过两层 for 循环，遍历整型数组中的每个元组，并进

```
//: Playground - noun: a place where people
   can play

import UIKit

func sortNumberArray(numbers : [Int]) ->
    [Int] {
    var sortedNumbers = numbers                     [23, 2, 12, 54, 87, 0, 4, 109, 20, 66]
    let count = numbers.count - 1                   9
    var temp = 0                                    0

    for i in 0...count {
        for j in 0 ... count {
            if sortedNumbers[i] <
                sortedNumbers[j] {
                temp = sortedNumbers[i]             (23 times)
                sortedNumbers[i] =                  (23 times)
                    sortedNumbers[j]
                sortedNumbers[j] = temp             (23 times)
            }
        }
    }

    return sortedNumbers                            [0, 2, 4, 12, 20, 23, 54, 66, 87, 109]
}

let numbers = [23,2,12,54,87,0,4,109,20,66]         [23, 2, 12, 54, 87, 0, 4, 109, 20, 66]
let sortedNumbers = sortNumberArray(numbers)        [0, 2, 4, 12, 20, 23, 54, 66, 87, 109]
```

图 26.20 整数排序的程序及运行结果

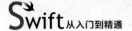

行两两比较,如果数组中第 i 个元素的值小于第 j 个元素的值,则交换两者元素的值。函数体的最后是返回排序后的整型数组。通过定义了一个含有 10 个元素的排序混乱的整型数组,将其作为参数,调用函数 sortNumberArray(),可以看到返回值为一个从大到小排序的数组。

【练习 21】 求 5×5 矩阵对角线之和

问题描述:给一个任意整型数组成的 5×5 矩阵,求组成对角线的整数的和。例如:

```
    1    2    3    4    5
    6    7    8    9   10
   11   12   13   14   15
   16   17   18   19   20
   21   22   23   24   25
```

对角线的整数为 1,7,13,19,25。

程序设计思路:求对角线之和的关键在于两点:第一,如何保存这个 5×5 矩阵;第二,怎么找到对角线的整数。第一个问题可以通过一个二维整型数组解决。在二维数组中,每个元素的下标即这个元素的位置信息。例如,[0][0]表示第一行第一列。第二个问题,通过观察可以发现,所有对角线上的整数的下标中,行数和列数都相等。例如,[2][2]就在对角线上。

求 5×5 矩阵对角线之和的程序及运行结果如图 26.21 所示。定义函数 sumOfDiagonal(),该函数含有一个参数 rectangle,为二维整型数组类型,用来接收外部传入的二维数组。函数体里定义了整型变量 sum,初始值为 0。定义了常量 rect,赋值为二维整型数组 rectangle。通过两层循环,取出行号和列号相等的元素,并进行累加,赋值给 sum。最后,打印计算对角线元素和的结果信息。

```
1   //: Playground - noun: a place where
2       people can play
3   import UIKit
4
5   func sumOfDiagonal(rectangle : [[Int]]){
6       var sum = 0                                     0
7       let rect = rectangle                            [[1, 2, 3, 4, 5], [6, 7, 8, 9, 10], [11,
8       for i in 0...4 {
9           for j in 0...4{
10              if i == j {
11                  sum += rect[i][j]                   (5 times)
12              }
13          }
14      }
15
16      print("The sum of diagonal is \         "The sum of diagonal is 65\n"
            (sum)")
17  }
18
19  let rectangle = [[1,2,3,4,5],                       [[1, 2, 3, 4, 5], [6, 7, 8, 9, 10], [11,
        [6,7,8,9,10],[11,12,13,14,15],[16,17,
        18,19,20],[21,22,23,24,25]]
20
21  sumOfDiagonal(rectangle)
```

图 26.21 求 5×5 矩阵对角线之和的程序及运行结果

【练习 22】 折半查找

问题描述：给定一个任意长度的整型数组，要求用折半查找法判断该数组中是否存在一个给定的整数。

程序设计思路：折半查找法首先假设表中元素是按升序排列，将表中间位置记录的数值与查找数值比较，如果两者相等，则查找成功；否则利用中间位置记录将表分成前、后两个子表，如果中间位置记录的数值大于查找数值，则进一步查找前一子表，否则进一步查找后一子表。重复以上过程，直到找到满足条件的记录，使查找成功，或直到子表不存在为止，此时查找不成功。本题中，在运用折半查找法之前需要将数组进行排序，然后再运用折半查找法。

折半查找的程序及运行结果如图 26.22 所示。定义函数 binarySearch()，该函数含有两个参数：整型数组 theArray，用来传入待查找元素的数组；整型数 theNumber 为待查找的元素。该函数有一个布尔型的返回值，用来表示是否查找到相同的元素。函数体中定义了 4 个变量：result 为布尔型，初始值为 false，作为返回值；整型变量 lo 表示当前比较的子数组的最低位序号，初始值为 0；整型变量 hi 表示当前子数组的最高位序号,初始值为 theArray 的长度减一;整型变量 mid 表示当前子数组的中间位置的序号。通过 while 循环,比较 lo 是否小于或等于 hi。在循环体中,首先通过 lo 和 hi 计算 mid,然后比较 theNumber 和 theArray[mid],如果 theNumber＜theArray[mid],将 hi 移动到 mid 前面,即 mid－1,然后进行下一轮比较;如果 theNumber＞theArray[mid],将 lo 移动到 mid 前面,即 mid＋1,然后进行下一轮比较;两种情况都不满足的唯一可能就是两者相等,说明匹配到相等的元素,返回值 result 赋值为 true,并终止循环。在函数体外,我们定义了一个乱序的整型数组,一共含有 10 个元素,先通过数组类型的方法 sort 对数组进行排序,然后将排过序的数组 sortedArray 作为参数,调用 binarySearch() 函数,变换不同的待查找整数,检查匹配结果,运行结果正确。

```
//: Playground - noun: a place where people can play

import UIKit

func binarySearch(theArray : [Int], theNumber : Int) ->
    Bool{
    var result = false                          (4 times)
    var lo = 0                                  (4 times)
    var hi = theArray.count - 1                 (4 times)
    var mid : Int
    while (lo <= hi) {
        mid = lo + ((hi - lo)/2)                (10 times)
        if theNumber < theArray[mid] {
            hi = mid - 1                        (4 times)
        }else if theNumber > theArray[mid] {
            lo = mid + 1                        (4 times)
        }else {
            result = true                       (2 times)
            print("result = true")              (2 times)
            break
        }
    }
    return result                               (4 times)
}

let theArray = [23,266,34,2,445,35,90,32,21,18]    [23, 266, 34, 2, 445, 35, 90, 32, 21, 18]
let sortedArray = theArray.sort()                  [2, 18, 21, 23, 32, 34, 35, 90, 266, 445]
binarySearch(sortedArray, theNumber: 32)           true
binarySearch(sortedArray, theNumber: 266)          true
binarySearch(theArray, theNumber: 25)              false
binarySearch(theArray, theNumber: 90)              false
```

图 26.22　折半查找的程序及运行结果

【练习 23】　围圈报数

问题描述:n 个人围成一圈,第一个人报数 1,第二个人报数 2,第三个人报数 3,下一个人再从 1 开始报数,每次有人报到 3,就将其剔除圈子,直到最后只剩下一个人,请问剩下的那个人在开始报数前,排在第几个?

程序设计思路:n 个人围成圈的报数问题可以通过一个布尔型一维数组建模,数组的下标表示每个人在报数前的位置,数组中每个元素的布尔值可以表示该人是否报到 3。初始时,数组中的每一个元素都应该为 true,表示都没有报到 3,每当有人报到 3,就将其标识为 false。每次报数的时候,只有为 true 的元素才能参加,每 3 个数就会少一个 true 的元素,如此往复,最后剩下的一个为 true 的元素,就是我们要求的元素,而这个元素的下标就是它在报数前的位置。

围圈报数的程序及运行结果如图 26.23 所示。定义函数 circle3(),这里的 3 表示数到 3 就将剔除一个元素,也可以将数到几作为函数的一个参数,这里假定是数到 3。函数有一个参数 num,为整型,用来传入一共有多少个人。函数体中定义了 4 个变量:布尔型数组 theArray,用来表示每个人是否报过 3,true 为没报过 3,false 为报过 3,初始化为一个布尔型空数组;整型变量 theRestNum 表示圈子里还剩下多少个人,初始值为 num;整型变量 count3 表示当前数到几了,初始值为 0;整型变量 index 表示数组

```
1   //: Playground - noun: a place where people can
       play
2
3   import UIKit
4
5   func circle3(num : Int){
6       var theArray  = [Bool]()                    []
7       var theRestNum = num                        12
8       var count3 = 0                              0
9       var index = 0                               0
10      for _ in 0...(num - 1) {
11          theArray.append(true)                   (12 times)
12      }
13      while theRestNum > 1 {
14          if theArray[index] == true {
15              count3 += 1                         (33 times)
16          }
17          if count3 == 3 {
18              count3 = 0                          (11 times)
19              theArray[index] = false             (11 times)
20              theRestNum = theRestNum - 1         (11 times)
21          }
22          index += 1                              (65 times)
23          if index == num {
24              index = 0                           (5 times)
25          }
26      }
27      for i in 0...(num-1) {
28          if theArray[i] == true {
29              print(i)                            "9\n"
30          }
31      }
32  }
33
34  circle3(12)
```

图 26.23 围圈报数的程序及运行结果

中元素的下标。首先,通过 for 循环初始化数组 theArray,初始化为长度为 num,每个元素值都为 true 的一个布尔型数组。然后通过 while 循环反复执行数数的操作,while 循环的停止条件是剩下的人数等于 1,即 theRestNum>1。在循环体内,首先从第一个数组元素开始判断是否报过 3,即是否等于 true,如果等于 true,说明没有报过 3,那么数 3 的变量 count3 加 1。然后再判断数 3 的变量是否等于 3,即是否已经数到 3 了,如果等于 3,就将数 3 的变量重置为 0,即从头开始数。同时将当前数组元素的值置为 false,表示已数过 3,退出圈子。另外,还要将圈子的长度减 1。然后,index 增加 1,表示报数的人移动到下一个人。如果 index 等于 num,说明遍历完一遍数组了,需要重头开始,所以 index 重置为 0。while 循环结束后,数组中只有一个元素的值为 true。通过 for 循环遍历数组,找到值为 true 的元素,将其位置打印出来。例如,12 个人的圈子,最后数数剩下的是位置(即数组中的下标)为 9 的人。

【练习 24】 求分数的和

问题描述:根据输入的 n 求分数的和,如果 n 为奇数,则求分数 1/1+1/3+1/5+…+1/n 的和;如果 n 为偶数,则求分数 1/2+1/4+1/6+…+1/n 的和。

程序设计思路:首先判断 n 的奇偶,可通过模 2 实现,若余数为 0,则为偶数,否则为奇数。其次,总结分数数列的规律,这里无论 n 为偶数,还是奇数,分数数列的规律都是一样的,即后一个分数的分母为前一个分数的分母加 2,一直加到 n。不同之处在于,第一个分数的分母不同,n 为奇数时,第一个分数的分母为 1,n 为偶数时,第一个分数的分母为 2。求出每个分数的规律,就可以通过循环求数列的和了。

求分数的和的程序及运行结果如图 26.24 所示。定义函数 fractionSum(),该函数含有一个整型参数 n,表示输入的数为 n。返回值为双精度浮点型,因为分数的值一般都是浮点型,返回值为分数数列的和的值。函数体里,定义变量 sum,初始值为 0.0,用来保存分数数列累加的和。

通过 if 语句判断 n 模 2 的值,若为 0,则说明 n 是偶数。由于偶数分母数列的第一项的分母为 2,所以赋值变量 i 为 2。通过 while 循环判断当前分母 i 的值是否小于或等于 n,如果等于 n,则执行下述操作,将分数 1/i 的值累加到 sum 中,然后 i 加 2,即得到下一个分数的分母。如果 i 的值大于 n,则跳出循环,结束累加计算。

```
//: Playground - noun: a place where
    people can play
import UIKit
func fractionSum(n : Int) -> Double {
    var sum = 0.0                              (4 times)
    if n % 2 == 0 {
        var i = 2                              (2 times)
        while i <= n {
            sum += 1/Double(i)                 (3 times)
            i += 2                             (3 times)
        }
    } else {
        var i = 1                              (2 times)
        while i <= n {
            sum += 1/Double(i)                 (3 times)
            i += 2                             (3 times)
        }
    }
    return sum                                 (4 times)
}
fractionSum(1)                                 1
fractionSum(2)                                 0.5
fractionSum(3)                                 1.333333333333333
fractionSum(4)                                 0.75
```

图 26.24　求分数的和的程序及运行结果

如果 n 模 2 的值不为 0，则说明 n 为奇数。由于奇数分母数列的第一项的分母为 1，所以赋值变量 i 为 1。通过 while 循环判断当前分母 i 的值是否小于或等于 n，如果等于 n，则执行下述操作，将分数 1/i 的值累加到 sum 中，然后 i 加 2，即得到下一个分数的分母。如果 i 的值大于 n，则跳出循环，结束累加计算。

计算结束后，返回 sum 的值给调用对象。

【练习 25】　字符串排序

问题描述：给定一个字符串数组，要求对其中的每个元素按照大小进行排序。

程序设计思路：前面介绍过一个整数排序的例子，这里换成了字符串，实际上在算法设计上是完全一样的，只不过整数换成字符串而已。在 Swift 中，可以直接比较字符串大小，而不需要将字符串中的字符取出，逐个进行比较，这些工作 Swift 已经全部做好了。

字符串排序的程序及运行结果如图 26.25 所示。定义函数 sortStrings()，该函数含有一个字符串数组类型的参数 strings，用来传入待排序的字符串数组。函数返回值同样为一个字符串数组类型，用来保存排好序的字符串数组。函数体内定义了字符串数组类型的变量 sortedString，用来存储排好序的字符串数组，初始值为 strings；定义了常量 length，用来表示字符串数组的长度；定义了字符串变量 temp，当交换两个数组元素的值时，用 temp 作为中转。通过两层 for 循环，比较两两相邻的数组元素大小，如果前一个元素小于后一个元素，则交换位置。第一层循环控制 i 是从 0 到 length－2，表示从数组的第一个元素开始一直到数组的倒数第二个元素为止，因为最后一个元素将由下一层循环取出进行比较，所以第一层循环只到倒数第二个元素即可。第二层循环控制 j 是从 i+1 开始到数组最后一个元素。起始值是从 i 之后的第一个元素开始，前面的元素都已经排好序了，不需要再取出进行比较。最后，返回排好序的字符串数组。通过实例可以看到，乱序的字符串数组已经按照由大到小的顺序排列好了。这里使用的算法比起整数排序的算法，效率高很多，关键点在于两层循环的控制取值范围不同，从而导致循环的次数不同。循环次数越少，排序速度越快。

```
//: Playground - noun: a place where people can
    play
import UIKit
func sortStrings(strings : [String]) -> [String] {
    var sortedString : [String]
    sortedString = strings
    let length = strings.count
    var temp = ""
    
    for i in 0...(length-2){
        for j in (i+1)...(length-1){
            if sortedString[i] < sortedString[j] {
                temp = sortedString[i]
                sortedString[i] = sortedString[j]
                sortedString[j] = temp
            }
        }
    }
    return sortedString
}
sortStrings(["how","many","roads","must","a","man",
    "walk","down"])
```

["how","many","roads","must","a","man","walk","down"]
8
""

(14 times)
(14 times)
(14 times)

["walk","roads","must","many","man","how","down","a"]

["walk","roads","must","many","man","how","down","a"]

图 26.25　字符串排序的程序及运行结果

【练习 26】　猴子分桃问题

　　问题描述：树底下有一堆桃子，有 n(n>1) 只猴子。第一只猴子把这堆桃子平均分为 n 份，多了一个，这只猴子把多的一个扔掉，拿走其中的一份。第二只猴子又把剩下的桃子平均分为 n 份，又多了一个，同样把它扔掉，并拿走其中的一份。其他猴子也是如此，直到第 n 只猴子还是分成 n 份，又多了一个。问一共有多少个桃子？

程序设计思路：首先 n 只猴子都做了同样的动作，说明这需要一个循环解决问题。其次，每只猴子都是将桃子分成 5 份，再扔掉一个，说明当时桃子数减 1 可以被 5 整除，这可以通过模 5 等于 0 表示。桃子数有可能取多个值，我们可以只考虑一个范围内符合条件的桃子数。所以，桃子数要用一个整型数组存储。

猴子分桃问题的程序及运行结果如图 26.26 所示。定义函数 monkeyDividePeach()，该函数有一个整型参数 numOfMonkey，用来表示猴子的数量。函数返回值为一个整型数组，用来保存桃子数的可能值。函数体中定义了 3 个变量，分别为 peachNum、flag、peachSum。peachNum 为整型变量，初始值为 0，用来表示当前的桃子数。flag 为布尔型变量，初始值为 true，用来表示当前尝试的数是否满足分桃的条件。peachSum 为整型数组变量，用来将满足分桃要求的桃子数添加到数组中。这里只考

```
//: Playground - noun: a place where people can play

import UIKit

func monkeyDividePeach(numOfMonkey : Int) -> [Int] {
    var peachNum = 0                                        0
    var flag : Bool
    var peachSum = [Int]()                                  []

    for i in 1...8000 {
        flag = true                                         (8000 times)
        peachNum = i                                        (8000 times)

        for count in 1...numOfMonkey {
            if (peachNum - 1) % 5 != 0 {
                flag = false                                (7994 times)
                break
            }
            peachNum = (peachNum - 1) / 5                   (1996 times)
            if peachNum == 1 {
                if count == numOfMonkey {
                    flag = true                             true
                } else {
                    flag = false                            (4 times)
                    break
                }
            }
        }

        if flag {
            print(i)                                        (2 times)
            peachSum.append(i)                              (2 times)
        }
    }

    return peachSum                                         [3,906, 7,031]
}

monkeyDividePeach(5)                                        [3,906, 7,031]
```

图 26.26 猴子分桃问题的程序及运行结果

虑桃子数的范围为 8000 以内。通过 for 循环，i 依次取值为 8000 以内的数，逐一进行尝试，看是否满足条件。在循环体中，首先置 flag 为 true，即假定当前的 i 为符合条件的桃子数。然后，将 i 赋值给 peachNum，即当前桃子数 peachNum 取值为 i。这里使用 peachNum，而不直接使用 i 有两个原因：一方面，i 是常量，值不能变化，而每只猴子分完桃子之后，当前桃子数都会变化；另一方面，peachNum 的可读性更好。第二层循环的执行次数由猴子数 numOfMonkey 决定，从而模拟每只猴子分桃的过程。循环体中，首先判断桃子数减 1 再模 5 是否等于 0，如果不等于 0，则置 flag 为 false，说明该数不符合条件，然后直接跳出第二层循环。如果等于 0，则继续往下执行。将当前桃子数减 1 再整除 5 得到的数赋值给 peachNum，作为下一只猴子的当前桃子数，在下一次循环时使用。最后还要判断一个特殊情况，即下一次分配的桃子数 peachNum 等于 1，同时 count 等于 numOfMonkey，即最后一只猴子已经分过了桃子，这种情况也满足条件，所以置 flag 为 true，否则置 flag 为 false。在第一层循环的最后部分，根据当前 flag 的值，决定是否将 i 添加到 peachNum 数组中。函数最后返回整个符合条件的数组。

【练习 27】 考试成绩统计

问题描述：统计学生的考试成绩，有 n 个学生，要求统计 n 个学生数学、语文、英语的平均分，以及打印出每个学生的各科成绩。

程序设计思路：首先需要定义一个学生类，该类中包含学生的个人信息和三门功课的成绩，同时要有打印学生基本信息的功能。n 个学生就是 n 个学生类的实例。统计 n 个学生的各科平均分，可以创建一个学生类的数组，将 n 个学生类实例添加到该数组中，然后问题就变成了对数组的循环操作问题。

考试成绩统计的程序及运行结果如图 26.27 所示。定义类 Student()，该类含有 4 个成员变量，分别为名字 name，字符串型；数学成绩 math，双精度浮点型；语文成绩 Chinese，双精度浮点型；英语成绩 English，双精度浮点型。类中定义了构造器 init，接收 4 个参数，为 4 个成员变量赋初始值。另外，类中还有一个 description 方法，用来打印学生的基本信息。

定义一个学生类的数组变量 stuArray，然后创建 3 个学生类实例 stu1、stu2、

```
5   class Student {
6       var name : String
7       var math : Double
8       var Chinese : Double
9       var English : Double
10      init(n : String, m : Double, c : Double, e : Double){
11          name = n
12          math = m
13          Chinese = c
14          English = e
15      }
16      func description(){
17          print("\(name)'s performance is : math \(math),        (3 times)
                Chinese \(Chinese), English \(English)")
18      }
19  }
20
21  var stuArray = [Student]()                                      []
22
23  let stu1 = Student(n: "Tom", m: 98.0, c: 86.0, e: 92.0)         Student
24  let stu2 = Student(n: "Sam", m: 99.0, c: 92.0, e: 96.0)         Student
25  let stu3 = Student(n: "Jerry", m: 73.0, c: 99.0, e: 80.0)       Student
26
27  stuArray.append(stu1)                                           [{name "Tom", math 98, Chinese 86, Eng
28  stuArray.append(stu2)                                           [{name "Tom", math 98, Chinese 86, Eng
29  stuArray.append(stu3)                                           [{name "Tom", math 98, Chinese 86, Engl
30
31  stu1.description()                                              Student
32  stu2.description()                                              Student
33  stu3.description()                                              Student
34
35  var sumOfMath = 0.0                                             0
36  var sumOfChinese = 0.0                                          0
37  var sumOfEnglish = 0.0                                          0
38  var countOfStu = stuArray.count                                 3
39
40  for i in 0...(countOfStu-1) {
41      sumOfMath += stuArray[i].math                               (3 times)
42      sumOfChinese += stuArray[i].Chinese                         (3 times)
43      sumOfEnglish += stuArray[i].English                         (3 times)
44  }
45
46  print("The average score of Math is " + String(sumOfMath/       "The average score of Math is 90.0\n"
        Double(countOfStu)))
47  print("The average score of Chinese is " + String              "The average score of Chinese is 92.333
        (sumOfChinese/Double(countOfStu)))
48  print("The average score of English is " + String              "The average score of English is 89.333
        (sumOfEnglish/Double(countOfStu)))
```

图 26.27 考试成绩统计的程序及运行结果

stu3，并调用其构造器进行初始化。将这 3 个实例添加到数组 stuArray 中，通过 Student 类中的方法 description() 打印每个学生的描述信息。

定义 3 个浮点型变量：数学成绩总分 sumOfMath、语文成绩总分 sumOfChinese、英语成绩总分 sumOfEnglish。定义一个整型变量 countOfStu，赋值为 stuArray 的长度。

通过 for 循环，遍历数组 stuArray 中的每个实例，取出其各科成绩，并进行累加，得到各科的总成绩。

最后，通过各科总成绩除以人数，得到各科的平均分，并打印出来。

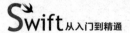

【练习 28】 子串出现的次数

问题描述：给定一个字符串，求该字符串中某子串出现的次数。

程序设计思路：Swift 中字符串类型有很多个方法，读者需要仔细参阅苹果的 Swift 官方文档。这里通过字符串类型的方法 componentsSeparatedByString() 求子字符串的出现次数。该方法有一个参数，即子字符串。该方法将子字符串作为一个分隔符，将字符串分成若干个子串，并存储到一个字符串数组中。这里存在一个关系，即字符串被分隔成 n 个子串，必然存在 n−1 个分隔符。由此，我们就可以计算子字符串出现的次数了。

子串出现的次数的程序及运行结果如图 26.28 所示。定义函数 countOfSubstring()，该函数含有两个字符串型参数：theString 表示待判断的字符串，theSubS 表示子字符串。函数返回值为整型，表示子字符串出现的次数。在函数体中，通过调用 theString 的方法 componentsSeparatedByString()，以字符串 theSubS 为分隔符，将 theString 分隔为若干个子字符串，并存储在字符串数组 strArray 中。依次打印数组 strArray 中的元素。将数组 strArray 的长度减 1 作为返回值。通过两个实例，可以验证该算法的正确性。

```
1  //: Playground - noun: a place where people can
       play
2
3  import UIKit
4
5  func countOfSubstring(theString : String,
       theSubS : String) -> Int {
6
7      let strArray = theString.
           componentsSeparatedByString(theSubS)       (2 times)
8      for i in 0...strArray.count - 1 {
9          print(strArray[i])                         (6 times)
10     }
11     return strArray.count - 1                      (2 times)
12 }
13
14 countOfSubstring("Hello world", theSubS: "ll")     1
15 countOfSubstring("Hello world", theSubS: "l")      3
```

```
He
o world
He
o wor
```

图 26.28 子串出现的次数的程序及运行结果

【练习 29】 数字加密问题

问题描述：某单位用公用电话线路传递数据。数据格式为 4 位整数。传递过程中进行加密,具体的加密算法：4 位数的每个数字先加上 5,再除以 10,得到的余数代替原来的数字,组成一个新的 4 位数。然后,该 4 位数的第 1 位和第 4 位进行交换,第 2 位和第 3 位进行交换。给定一个 4 位整数,要求进行加密。

程序设计思路：对一个 4 位整数加密,首先需要将 4 个数字分离出来,进行加密处理后再合并成一个新的 4 位整数。

数字加密问题的程序及运行结果如图 26.29 所示。定义函数 encryptNumbers(),该函数含有一个整型参数 originNumber,为待加密的 4 位整数。函数的返回值为整型,是加密后的 4 位整数。函数体内定义了 3 个变量：整型数组变量 bitArray,用来存每位上的数字；整型变量 temp,用作两个数字交换时的缓冲量；整型变量 encrypted 用来存储加密后的 4 位整数。

```
//: Playground - noun: a place where people can play
import UIKit
func encryptNumbers(originNumber : Int) -> Int {        (3 times)
    var bitArray = [Int]()
    var temp : Int
    var encrypted : Int
    bitArray.append(originNumber / 1000)                (3 times)
    bitArray.append(originNumber / 100 % 10)            (3 times)
    bitArray.append(originNumber / 10 % 10)             (3 times)
    bitArray.append(originNumber % 10)                  (3 times)

    for i in 0...3 {
        bitArray[i] += 5
        bitArray[i] = bitArray[i] % 10                  (12 times)
    }

    temp = bitArray[0]                                  (3 times)
    bitArray[0] = bitArray[3]                           (3 times)
    bitArray[3] = temp                                  (3 times)

    temp = bitArray[1]                                  (3 times)
    bitArray[1] = bitArray[2]                           (3 times)
    bitArray[2] = temp                                  (3 times)

    encrypted = bitArray[0]*1000 + bitArray[1]*100 +    (3 times)
        bitArray[2]*10 + bitArray[3]

    return encrypted                                    (3 times)
}

encryptNumbers(2678)                                    3217
encryptNumbers(5678)                                    3210
encryptNumbers(2936)                                    1847
```

图 26.29　数字加密问题的程序及运行结果

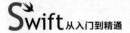

首先，依次将 4 位整数的千、百、十、个位数字提取出来，并依次存到数组 bitArray 中。通过 for 循环对数组 bitArray 中的每个元素进行加 5 操作，然后再模 10，最后保存回原来的位置。循环结束后，再将数组 bitArray 中的第 1 个元素 bitArray[0] 和第 4 个元素 bitArray[3] 的值互换。同样，互换第 2 个元素 bitArray[1] 和第 3 个元素 bitArray[2]。此时，数组 bitArray 中的元素就是加密后的 4 位整数的千、百、十、个位上的 4 个数字。将 bitArray 中的元素依次取出来，分别乘以 1000、100、10、1，然后再相加，即得到加密后的 4 位整数 encrypted。

【练习 30】 被 9 整除问题

问题描述：写一个函数，判断任意一个整数能够被另一个任意整数整除的次数。

程序设计思路：函数有两个参数：一个为被除数；另一个为除数。函数的返回值为被整除的次数。一个数能被另一个数整除，也就是说，一个数被另一个数取模后，余数为零，可以用这种方式判断一个数是否被另一个数整除。可以通过循环实现一个数被另一个数反复整除，并记录整除的次数，一旦不能被整除，则退出循环。

被 9 整除问题的程序及运行结果如图 26.30 所示。定义函数 dividedWithoutRemainder()，该函数有两个整型参数：number 表示被除数，divider 表示除数。函数

```
//: Playground - noun: a place where people can play

import UIKit

func dividedWithoutRemainder(number : Int, divider : Int) -> Int {
    var count = 0                                                   (2 times)
    var theNumber = number                                          (2 times)
    while theNumber % divider == 0 {
        theNumber = theNumber / divider                             (5 times)
        count += 1                                                  (5 times)
        if theNumber == 0 {
            break
        }
    }
    print("The number \(number) can be divided by \(divider) without
        remainder for \(count) times")                              (2 times)
    return count                                                    (2 times)
}

dividedWithoutRemainder(729, divider: 9)                            3
dividedWithoutRemainder(800, divider: 4)                            2
```

The number 729 can be divided by 9 without remainder for 3 times
The number 800 can be divided by 4 without remainder for 2 times

图 26.30 被 9 整除问题的程序及运行结果

返回值为整型,表示 number 可以被 divider 整除的次数。函数体中定义了两个整型变量:count 表示目前被整除的次数,初始值为 0;theNumber 表示当前被除数的值。因为 theNumber 每次被整除后,值都会变化,所以要定义一个变量表示,而不能直接使用 number,这里 number 为常量。通过 while 循环反复用 divider 整除 theNumber。循环的入口条件为当前 theNumber 被 divider 取模后为 0,即 theNumber 可以被 divider 整除。循环体中,theNumber 除以 divider,结果保存到 theNumber 中,然后整除的次数 count 加 1。最后,判断 theNumber 是否为 0,如果为 0,就直接跳出循环,否则会进入死循环,因为 0 和任何数取模都为 0。循环结束后打印相关信息,并返回整除的次数。

图书资源支持

感谢您一直以来对清华版图书的支持和爱护。为了配合本书的使用,本书提供配套的资源,有需求的读者请扫描下方的"书圈"微信公众号二维码,在图书专区下载,也可以拨打电话或发送电子邮件咨询。

如果您在使用本书的过程中遇到了什么问题,或者有相关图书出版计划,也请您发邮件告诉我们,以便我们更好地为您服务。

我们的联系方式:

地　　址: 北京市海淀区双清路学研大厦 A 座 701

邮　　编: 100084

电　　话: 010-83470236　010-83470237

资源下载: http://www.tup.com.cn

客服邮箱: tupjsj@vip.163.com

QQ: 2301891038(请写明您的单位和姓名)

用微信扫一扫右边的二维码,即可关注清华大学出版社公众号"书圈"。

资源下载、样书申请

书圈

扫一扫,获取最新目录

课程直播